Energy Needs
and the
Environment

SYMPOSIUM ON ENERGY,
"THE ENVIRONMENT, AND
EDUCATION, UNIVERSITY
OF ARIZONA, 1971

Energy Needs
and the
Environment

editors

ROBERT L. SEALE
RAYMOND A. SIERKA

THE UNIVERSITY OF ARIZONA PRESS
Tucson, Arizona

About the Editors

ROBERT L. SEALE since 1961 has been a member of the Nuclear Engineering faculty at the University of Arizona, becoming head of the department in 1969. He received his Bachelor of Science degree in physics from the University of Houston in 1947, and his Ph.D. from the University of Texas in 1953. His industrial experience has included employment as a geophysicist for the Gulf Oil Research and Development Company and, during the period from 1953 to 1961, employment in various nuclear engineering capacities with General Dynamics, Fort Worth. His research includes neutron slowing-down and nuclear kinetics problems. At the University of Arizona he has been active in the nuclear safety and power systems operations and design areas.

RAYMOND A. SIERKA joined the University of Arizona faculty in 1969 and became associate professor of civil engineering in 1970, teaching in the graduate sanitary engineering program and conducting research in the areas of adsorption phenomena and high-temperature waste treatment. Sierka's undergraduate studies in chemical engineering were completed at the University of Pittsburgh in 1960 when he became a process development engineer in the chemical industry, concentrating on industrial waste treatment activities. In 1969 he received the Ph.D. degree from the University of Oklahoma.

THE UNIVERSITY OF ARIZONA PRESS

Preface

The enormity and complexity of the interrelated problems of energy and the environment transcend traditional political and scientific boundaries. Therefore a group of representatives from government, the energy industry, and academia were invited to express their views and proffer suggestions concerning these problems and their solution. We feel the presentation of diverse ideas to this fundamental problem is imperative in view of the energy crises facing the world today. It is our hope that the data and opinions expressed in this volume will be read and considered thoughtfully by interested citizens and scientists alike, and utilized in their decision-making processes in the future. This volume evolved from the Symposium on *Energy, the Environment, and Education* held at the University of Arizona in the spring of 1971.

We are grateful to the Arizona Public Service Company, Tucson Gas and Electric Company, and the Salt River Project for their interest, encouragement, and financial support which helped to make this gathering possible. The assistance of the Arizona Atomic Energy Commission and of its Director, Don Gilbert, is also acknowledged. In addition, we are deeply indebted to George W. Howard, Roy O. McCauldin, M. R. Llewllyn, Lynn E. Weaver, and Richard Kassander, who served as session chairmen; to all of the speakers who so willingly gave of their time and expertise to make this meeting a success; to Henry Perkins and Roy Post of the College of Engineering who participated in the initial organization and planning sessions and arranged for various speakers; and to James Kendall who helped in the initial planning of this volume. Finally, we wish to thank the University of Arizona Press and, in particular, Marshall Townsend, for its help in the preparation and publication of this book.

Robert L. Seale
Raymond A. Sierka

CONTENTS

TABLES

ILLUSTRATIONS

1. The Challenge of Energy Generation

John A. Carver
Commissioner, Federal Power Commission

We pride ourselves in having mastered the social as well as the physical sciences, but I am far from confident that we understand the ecological or environmental revolution which we are now experiencing.

Idle speculation triggered by reading about the French Revolution recently caused me to think of parallels. The French Revolution has been described as "vast, shapeless, anonymous but irresistible." The ideas of Rousseau's *Social Contract* captured the ordinary Frenchman and found their way into the Declaration of the Rights of Man and Citizen in 1789.

This legacy suggests the proposals of the Sierra Club and other environmentalists aimed at legislating a Bill of Environmental Rights, and I have raised the question whether Rachel Carson's *Silent Spring* or Stewart Udall's *Quiet Crisis* is the appropriate book to compare with *Social Contract.*

Louis XVI's unshakable dedication to the divine right of kings was the force that had to be broken by revolution. Is it possible that an unshakable dedication to a constantly expanding economy or a constantly growing GNP faces a similar fate?

GROWING ENVIRONMENTAL CONCERN

Was the battle over the SST a skirmish in a developing larger struggle? It is certain that this controversy tells us a great deal about how pervasive and powerful the environmental issue has become and how these new

1

pressures originated. They are coming from the "vast, shapeless, anonymous but irresistible" American public. To those of us who live and work in Washington, D.C., it was instructive to see a struggle wherein organized labor, the national administration, and powerful business and regional lobbies were arrayed in a losing struggle against . . . against whom?

That is the exact point, and the source of my parallel with another kind of revolution in another era—it is not so much that the environmental movement has no leaders as that it needs no leaders. In the late 1960s, the movement which we used to call the Conservation Movement went "critical," and the reaction since then has been self-sustaining.

A newspaper man representing hundreds of papers told me that his papers would take anything he gave them on the environment. Environment fills the pages of the Congressional Record, of the journals of the professional societies, of the publications of the universities. The environment was a topic in the State of the Union Message and the subject of a separate presidential message in 1971. We have a National Council on Environmental Quality, an Environmental Protection Agency, and environmental subcommittees on approximately a half-dozen congressional standing committees.

The American passion for denigration by epithet has given us some new terms, such as "enviromaniac." The point may be lost that it is not the movement but the extremism which is attacked.

It is not unprecedented for a doomsday mania to spread widely, but something more than an unreasoning fear about the world's imminent end seems to be evident. Consciousness of the forces which alter man's physical environment may permeate the populace, but it is against institutions, not fate, that the people cry.

Along with news stories and features about the environment, pollution, and ecology, your daily newspaper on any given day is likely to be talking about a national energy crisis, a local or regional power shortage, or the problems of getting new or additional natural gas service for a community. Increasingly, these accounts reflect some direct tie to the economic welfare of the community: a housing project stalled for lack of gas, a plant shut down or prevented from expanding for lack of industrial fuel, or possibly a general commercial slowdown resulting from voltage reductions.

Equally commonplace is the observation that ours is an energy-based society which has increased its requirements in geometric terms since the middle of the last century. Aiming for a more sophisticated level, some of us have repeated the point that the advancement of modern "civilization" can be plotted on a curve that parallels that of energy consumption.

HENRY ADAMS FOREWARNS OF CONFLICT

This concept seemed to me to be a rather new insight until Lewis Mumford's most recent work of socio-technical commentary, *The Pentagon of Power,* rather burst the bubble. In his fourth and presumably final volume of the series, called *The Myth of the Machine,* Mumford records that his noted predecessor as social historian of the Victorian generation, Mr. Henry Adams, made the same observation as early as 1910.

It is worthy of note that on this occasion Adams was addressing himself to his colleagues. Some fifteen years after his retirement as president of the American Historical Association he prepared quite a long essay and distributed it to the Association under the modest title, "A Letter to American Teachers of History." Its objective was to trace the energy-related discoveries in physical science and to forewarn his colleagues of their responsibilities for the proper interpretation of these developments. That burden consisted principally of preparing society for the institutional changes and philosophical revisions necessary for existence in a physical environment which would be destined to change at an ever accelerating rate.

Adams apparently was engaging in a bit of interdisciplinary warfare in the academic world of his day. In Adams' essay, one feels, rather than reads, a concern for the conflict in roles between the physical sciences and social interpretation in the developing university structure. In the end, however, he hopes for the birth of a latter-day Newton who can reconcile the conflicting theories of degradation versus evolution—that is, of man as a waster of energy through inefficient conversion versus the ever-upward-progression view that had become attached to Darwinian biology.

It would be interesting to research the reception which this message received in both camps. But I have found no account which reveals this information. Certainly it must have had a restricted audience, for Adams was talking physics to historians and social dynamics to scientists—not a guaranteed formula for best-seller status. Equally clearly, however, he diagrammed the next six decades of America's developing dilemma with uncanny accuracy.

Over those six decades, science and technology by and large have remained locked in their laboratories insofar as social awareness is concerned. Institutions and philosophy did not modify or adjust or attempt to temper the impact of science. This is not a unique development; social institutions have traditionally followed a lag pattern in adjusting to change, sometimes measured in centuries. Some of the basic concepts of our real property law, for example, are rooted in the compelling necessities of the feudal system.

The critical difference is that the collapse of feudalism had no direct impact on the earth's resource inventory, nor did it involve the dissipation of unharnessed fossil or nuclear heat into the water or atmospheric shell of this planet. And so, in addition to the energy crisis that poses hazards for our economy, we have also an environmental crisis that poses hazards for our air, water, and food supply.

Having profited very little from Henry Adams' prescience of 1910, the question before our house today is whether we have learned very much in the hard school of experience. Are we facing the hard realities of environmental revolution in realistic fashion? What is the role of education?

Simple and categorical answers to these questions would necessarily be somewhat subjective and not very persuasive. There are, however, some indicators available that suggest possible answers.

ENVIRONMENTAL PROTECTION AND CONTROL

Increasingly over the past decade, and especially in the last half of that period, significant legislation has been enacted to protect various aspects of our natural heritage against further degradation. Water pollution control became an early national objective and its institutional structure has been expanded by successive amendments. Air quality and solid waste disposal are more recent sources of concern. Finally, in 1969–70, the National Environmental Protection Act provided an umbrella-type statement of national policy, requiring that each and every project involving government participation or approval be subjected to careful environmental impact scrutiny.

Each of these measures has created, in one degree or another, a constraint on the production, consumption, or transmission of energy resources or supplies. As a very minimum, they have delayed the process of keeping pace with energy consumption and demand. In other situations, certain fuels with high pollution content have been replaced by less objectionable energy sources.

Continuing on this course, on 8 February 1971 the president sent his environmental message to Congress, outlining a sixteen-point program to "maintain the initiative so vigorously begun in our shared campaign to save and enhance our surroundings." Seven of the sixteen program elements will or can have direct impact on energy production and use:

1. Use of the tax structure to reduce sulfur oxide emissions
2. Strengthened standards and enforcement authority for water quality control
3. Controls on ocean dumping
4. Development of a land use policy which would necessarily affect mineral production, power plant siting, and transmission line location

5. Expansion of the wilderness areas system
6. Advance governmental approval of power plant siting and transmission lines
7. Environmental controls over both surface and subsurface mining operations

Clearly these objectives are laudable. But equally clearly they can be achieved only at a price, both in public outlays for facilities and regulatory expense and in increased cost of goods and services which are dependent on continued availability of economical sources of energy.

The less tangible and direct impact which environmental constraints must inevitably exert on our general economic capability has not been calculated or even faced realistically. To those who are totally committed to environmental reform above all other values or objectives, this presents little difficulty. We must do without additional development of energy supplies. If this means indefinite or even permanent abandonment of economic growth, that is the price for retention of a planet capable of sustaining existing life forms.

M. King Hubbert of the United States Geological Survey has concluded that exhaustion of energy resources will eventually compel an end to the pattern of exponential growth; the alternatives are a levelling off at a high plateau or the completion of the characteristic curve back down to primitive existence. In closing the chapter on energy resources for *Resources and Man,* published for the National Academy of Science, Hubbert says:

It is paradoxical that although the forthcoming period of nongrowth poses no insuperable physical or biological problems, it will entail a fundamental revision of those aspects of our current economic and social thinking which stem from the assumption that the growth rates which have characterized this temporary period can be permanent.

Nearly sixty years apart, therefore, the modern geophysicist and the Victorian student of history meet on remarkably common ground.

The great difficulty in dealing with nongrowth is whether we define this concept as a solution or a problem. Certainly nongrowth has not yet been accepted as national policy. Indeed the most current indicator seems diametrically opposed to that concept. One week prior to his 1971 environmental message, the president sent to Congress his annual Economic Report, accompanied by the report of his Council of Economic Advisors. The keynote of the president's relatively brief summation is found in this conclusion:

In the record of progress toward that new prosperity, I am convinced that the economic historians of the future will regard 1970 as a necessarily difficult year of turnaround—but a year that set the stage for strong and orderly expansion.

The basis for this optimistic, growth-oriented forecast evidently stems from a full-employment budget previously submitted and the game plan set forth in the accompanying CEA report. The key elements or underpinning of that plan include the figure of 1.065 trillion dollars as an appropriate intermediate target for the GNP in 1971 and a gradual reduction in the unemployment rate from the 5.8 percent that prevailed at the end of 1970 to a more acceptable level of 4.5 percent by early 1972. It is clear that these target figures entail a 9.1 percent increase in gross national product in a year immediately following one in which economic growth was about half that rate.

The number of new jobs needed to meet these goals is not clearly stated, but a reduction in the rate of unemployment by roughly one-fourth, plus additions to the workforce through population growth and reduction in military strength, suggests that something in excess of 1.5 million new civilian jobs will be required in the relatively near term. It is axiomatic that the economic expansion required to support these new jobs cannot be achieved without a comparable increase in prime mover energy supply. Yet our national policy proposals for environmental reform make no reference to their impact on economic objectives.

In fairness, the CEA did not ignore the environmental issue; they talked about the cost aspects of pollution abatement and the need for a new set of rules to govern use of such common property as air and water.

The overall economic cost of pollution control has been recognized for some time. To some extent, marginal products or services may disappear as they are required to absorb their share of environmental protection costs. In as crucial an area as the production and use of basic energy resources, however, the problem cannot be easily resolved. The plain facts are that a point of no return must eventually be reached. If environmental necessity dictates zero growth, then you cannot increase GNP and you cannot enjoy ever-expanding employment and economic opportunity. The area for possible compromise grows narrower every year.

Here perhaps is the most serious issue of public policy facing this nation. I cannot dictate where the line should be drawn, nor should I as an individual. This is a decision the public should make and convey to its political leadership. However, we are still trying to "have it both ways" insofar as public policy proposals are concerned and this will not work. It can only result in confusion.

If our country does not continue to have sufficient electric power to meet its demands for a clean secondary form of energy, the environment will surely suffer beyond the rectification power of legislation framed for an environmental goal. Environmental protection is a crucial and vital aspect of the public interest, but to give environmental protection agencies a special and controlling role with respect to such matters as electric power generation is to invite disaster broader than ecological disaster. A failure of energy supply will be disruptive of all of our cleanup plans, because without energy to fuel the economy we will not achieve the dividends for the public good which we expect from its healthy operation.

We have defined, in the context of Henry Adams' and King Hubbert's terms, the necessity for a national social policy which will come to grips with the implications of an ultimate exhaustion of energy supply and particularly with our incapability of maintaining exponential growth rates indefinitely.

The physical facts must be understood, which calls for a higher quality of political leadership than we are generally seeing. Our political and economic system is complex, but it is undeniably a fact that inhibiting the use of energy, or the rate of conversion of primary to secondary energy, will accomplish the result, although certainly not necessarily the solution, of nongrowth. This simplistic fact has led to recommendations—often by those in Congress and in responsible executive positions who should be more aware—that politically unresponsible (in the sense of owing appointment to no electorate) agencies of government be authorized to close the valves and pull the switches in the name of protecting the environment.

In such a direction lies nongrowth, but also chaos, loss of confidence in the processes of government, and economic disruption with the potential of violence.

I am indebted again to Lewis Mumford for another reference to Adams, who in 1905 wrote as follows:

> The assumption of unity, which was the mark of human thought in the Middle Ages, has yielded very slowly to the proofs of complexity. The stupor of science before radium is a proof of it. Yet it is quite sure, according to my score of ratios and curves that, at the accelerated rate of progression since 1600, it will not need another century or half-century to turn thought upside down. Law in that case would disappear as theory or a priori principle and give place to force. Morality would become police. Explosives would reach cosmic violence. Disintegration would overcome integration.

Mumford states that in this passage Adams foresaw the social consequences of increasing physical power without commensurate intellectual insight, moral discipline, social awareness, and responsible political direction. These values summarize what Henry Adams asked of the teachers of history he addressed in his "letter" of 1910. It is a good summary, also, of what we ought to be able to expect of our educational system.

2. Energy Resources for Power Production

M. King Hubbert
United States Geological Survey

An adequate appraisal of the earth's energy resources appropriate for power production requires both an inventory of the earth's energetic processes as of a given time and a review of some of the more important changes that occur as functions of time. For the latter purpose, two time scales are essential: a geological time scale during which the earth's largest stores of energy have accumulated and a human time scale during which those accumulations are dissipated.

The world's energy resources suitable for power production are of two classes: (1) various channels of the continuous energy flux from extraterrestrial sources and from the earth's interior, and (2) chemical, thermal, and nuclear energy stored in the outer part of the lithosphere and in the oceans.

The energy flux (fig. 2.1) comprises an approximate steady state whereby the energy inputs into the earth's surface environment from these various sources undergo a series of thermodynamically irreversible degradations, the end product of which is heat at the lowest temperature of the environment. This then leaves the earth by long-wavelength thermal radiation. Accompanying this energy flux, the material constituents of the earth's surface undergo intermittent or continuous circulation.

A small fraction of matter on the earth comprises the biomass of the

NOTE: Publication authorized by the Director, United States Geological Survey. Revised from M. King Hubbert, "Energy Resources for Power Production," in *Environmental Aspects of Nuclear Power Stations* (Vienna, 1971): 13–43.

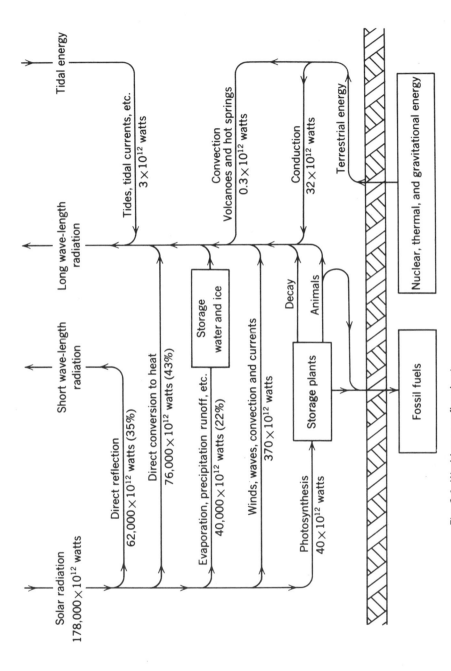

Fig. 2.1 World energy flowsheet.
SOURCE: Modified from Hubbert 1962, *Energy Resources*, fig. 1, p. 2.

10

earth's plant and animal kingdoms. A minute fraction of the incident solar radiation is captured by the plant leaves and is stored chemically by the process of photosynthesis whereby such inorganic materials as H_2O, CO_2, and mineral salts are synthesized into complex organic compounds. This provides the energy essential for the sustenance of the entire plant and animal kingdoms.

During the last 700 million years of geologic time, a small fraction of the remains of contemporary plants and animals have accumulated in oxygen-free environments, such as swamps, under conditions of incomplete decay, and have become buried under accumulations of sedimentary muds and sands. These, by subsequent chemical changes, have gradually become metamorphosed into the earth's present stores of the fossil fuels, coal, petroleum, and related substances.

During the million or two years since the human species evolved from its prehuman ancestors, this species has had a record of continuous achievement in the capture and control of ever-larger fractions of the contemporary energy. However, until about 800 years ago this consisted solely of the utilization of energy from one or another channel of solar-energy flux.

Man's dependence upon contemporary solar energy was broken when the energy from the fossil fuels was first tapped about the twelfth century A.D. with the beginning of coal mining in northeast England. Until the development of the first successful steam engine, that of Newcomen about 1712, the uses of coal were limited to fuel. Since the advent of the steam engine, the use of coal as a source of mechanical, and later electrical, power has proliferated. Continuous exploitation of the second major source of energy from the fossil fuels, that from petroleum, had its beginning in Romania in 1857 and two years later in the United States.

GROWTH OF POWER PRODUCTION

Although the use of water power and wind power for industrial purposes began more than 2,000 years ago, and the use of steam power with the development of the Newcomen engine about 1712, power production did not reach a significant magnitude in terms of current rates until approximately a century ago. The rapid growth in power production subsequently has been made possible principally by the development of electrical means for the generation and distribution of power from central power stations, and by the proliferation of much smaller mobile power plants for use by motor vehicles, aeroplanes, locomotives, and ships, based principally on the use of petroleum for fuel.

Electric power capacity

The world electrical generating capacity for the period 1955–1967 is shown in figure 2.2. During that period, this capacity increased at an average rate of about 8 percent per year, or with a doubling period of 8.7 years. Annual data for the years before 1955 have not been found, but in view of the fact that the first central electric power station was installed as recently as 1882, the power capacity by 1900 must still have been very small, and the growth between 1900 and 1955 must have been approximately as shown by the dashed curve in figure 2.2.

Mobile power units

Only limited statistical data on the world growth of mobile power units are available. However, from the *Statistical Yearbook 1968* of the United Nations, data on the growth rates of passenger automobiles, of kilometers flown on scheduled services of civil aviation, and of registered tonnages of merchant fleets within the last two decades have been obtained. These are shown in table 2.1 from which it will be seen that, for the period 1953–1967,

TABLE 2.1
World Growth Rates of Mobile Power Units

Passenger Automobiles					
Year	Number (thousands)	Time Interval (years)	Ratio of Increase	Rate of Increase (percent/yr.)	Doubling Time (years)
1953	62,910				
1967	160,240	14	2.55	6.69	10.4
Civil Aviation Schedule Flights					
	Kilometers flown (10^6 km.)				
1953	1,940				
1967	5,290	14	2.73	7.17	9.7
Merchant Shipping					
	Gross registered tons (thousands)				
1953	93,352				
1968	194,152	15	2.08	4.88	14.2

SOURCE: United Nations Statistical Yearbook 1968.

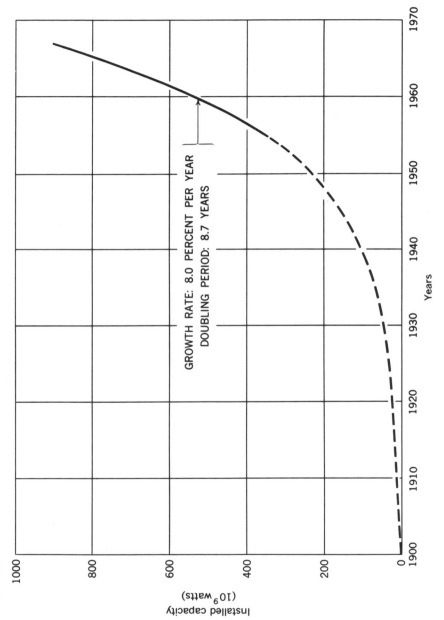

Fig. 2.2 World electrical generating capacity.
SOURCE: Hubbert 1971, "Energy Resources for Power Production" in *Environmental Aspects of Nuclear Power Stations*, fig. 2, p. 15.

the number of passenger automobiles increased at a rate of 6.69 percent per year with a doubling period of 10.4 years, and the distance flown on scheduled services of civil aviation increased at the rate of 7.17 percent per year, doubling every 9.7 years. For the period 1953 to 1968, gross tonnage of merchant shipping increased at a rate of 4.88 percent per year with a doubling period of 14.2 years.

From these data, it is evident that the combined world power capacity of both stationary electric power stations and mobile power units is increasing exponentially at such a rate as to double about once per decade.

GROWTH OF INDUSTRIAL ENERGY PRODUCTION

Coal

Although the mining of coal began some 800 years ago, in comparison with present magnitudes of the rates of production those prior to the year 1800 were almost negligible. Before 1860, reliable statistics on the world rate of production are difficult to assemble. For each year since 1860, the world annual production of coal and lignite is shown in figure 2.3. The same data, plotted semi-logarithmically, are shown in figure 2.4. From the latter, it will be seen that from 1860 to 1914, the beginning of World War I, production of coal and lignite increased at a steady exponential rate of 4.41 percent per year, with a doubling period of 16.1 years. During the following thirty-year period from 1914 to 1944, encompassing two world wars and an intervening depression, the growth rate slowed to an average rate of only 0.75 percent per year with a doubling period of ninety-three years. Since 1944, a rapid exponential growth at a rate of 3.56 percent per year, with a doubling period of 19.8 years, has been resumed.

Crude oil

Figure 2.5 shows the growth of the rate of the world production of crude oil since 1880, and in figure 2.6 the same data are plotted semi-logarithmically. From 1890 to the present, with only minor variations, world crude-oil production has increased at a steady exponential rate, averaging about 6.94 percent per year and doubling about every ten years.

Total energy from coal and crude oil

Finally, in figure 2.7, the energy production, expressed in the common unit of 10^{12} thermal kilowatt-hours per year, of coal and lignite plus crude oil from 1860 to 1965, is shown. Until after 1890 the energy contribution from petroleum, as compared with that of coal, was negligible. Subsequently, because the growth rate of crude-oil production is much larger

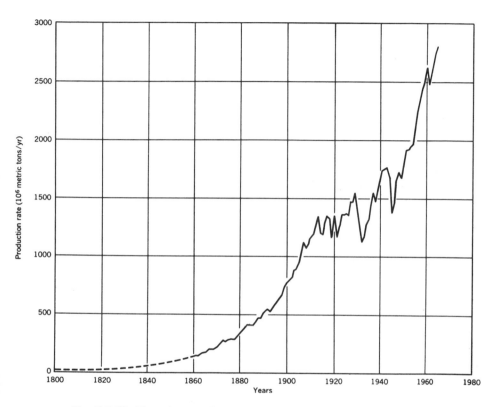

Fig. 2.3 World production of coal and lignite (arithmetic scale).
SOURCE: Hubbert 1969, *Resources and Man,* fig. 8.1, p. 161.

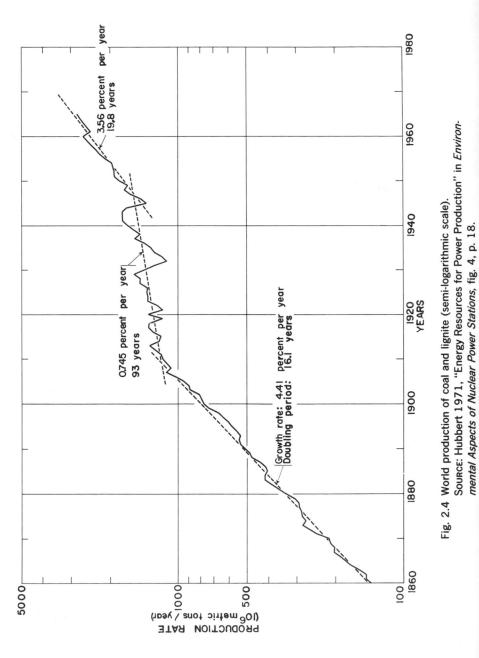

Fig. 2.4 World production of coal and lignite (semi-logarithmic scale).
Source: Hubbert 1971, "Energy Resources for Power Production" in *Environmental Aspects of Nuclear Power Stations*, fig. 4, p. 18.

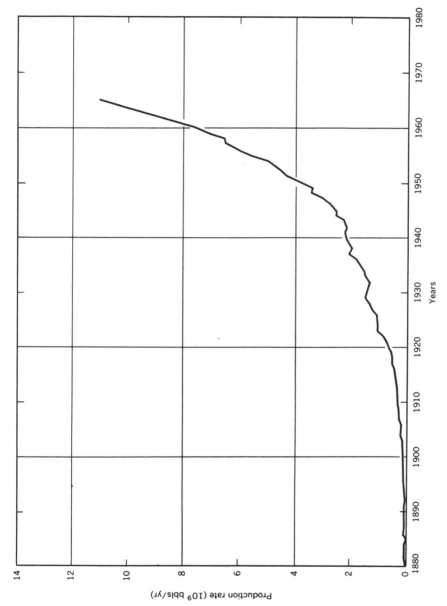

Fig. 2.5 World production of crude oil (arithmetic scale).
SOURCE: Hubbert 1969, *Resources and Man*, fig. 8.2, p. 162.

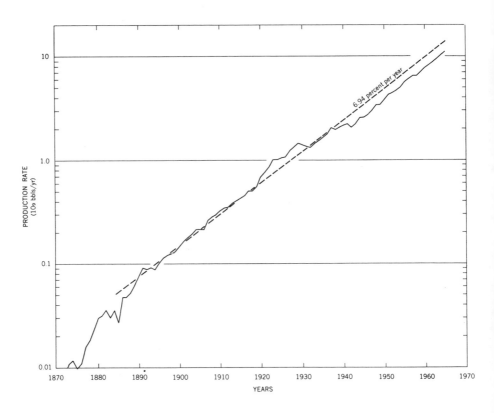

Fig 2.6 World production of crude oil (semi-logarithmic scale).
SOURCE: Hubbert 1971, "Energy Resources for Power Production" in *Environmental Aspects of Nuclear Power Stations,* fig. 6, p. 19.

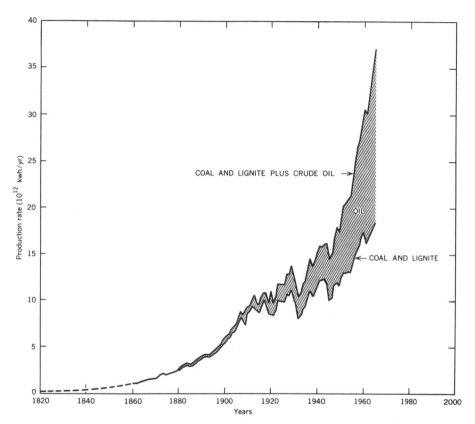

Fig. 2.7 World production of energy from coal and lignite plus crude oil.
SOURCE: Hubbert 1969, *Resources and Man,* fig. 8.3, p. 163.

than that of coal and lignite, its contribution has steadily increased until by 1965 the two were approximately equal. If the energy from natural gas were to be added to that of petroleum, the two together would account for about 60 percent of the total.

Although this is not shown in figure 2.7, next to the fossil fuels, the largest source of energy for power production in 1965 was water power. The hydroelectric energy produced in 1965 amounted to 0.898×10^{12} kilowatt-hours. This is but 2.4 percent of the 37×10^{12} thermal kilowatt-hours contributed during that year by coal, lignite, and crude oil. Hence, up to the present time, although the contribution of water power is significant, the predominant source of energy for industrial uses is still the fossil fuels.

FUTURE OF THE FOSSIL FUELS

Principle of prediction

A question of great importance with respect to the fossil fuels is approximately how long can they be depended upon as major sources of energy? A very satisfactory answer to this question for any given class of fuel can be obtained by use of the principle illustrated in figure 2.8. If the curve of the production rate of such a fuel is plotted on an arithmetic scale, as has been done for coal and oil in figures 2.3 and 2.5, this curve during the entire cycle of production must have the following properties: It must begin at zero, increase until it passes one or more maxima, and finally decline to zero. As illustrated by figure 2.8, the area beneath such a curve from the beginning up to any given time will be proportional to the cumulative production up to that time. For the complete cycle of production, the area under the curve cannot exceed the extractable amount of the resource initially present. Hence when we already have the production history up to the present, if we can estimate by geological or other means the approximate magnitude of the resource that remains, we can extrapolate the production curve for the rest of the cycle, subject to the foregoing restriction. While a multiplicity of such curves can be drawn, all such curves that are also consistent with the technology of production tend to have a strong family resemblance to one another.

World coal

Because coal occurs in sedimentary basins in strata that are commonly of large areal extent and that frequently crop out on the surface, reasonably reliable estimates of the coal resources of a given basin can be made by surface geological mapping supplemented by a comparatively small number of widely spaced drill holes. The most recent estimates of the initial quanti-

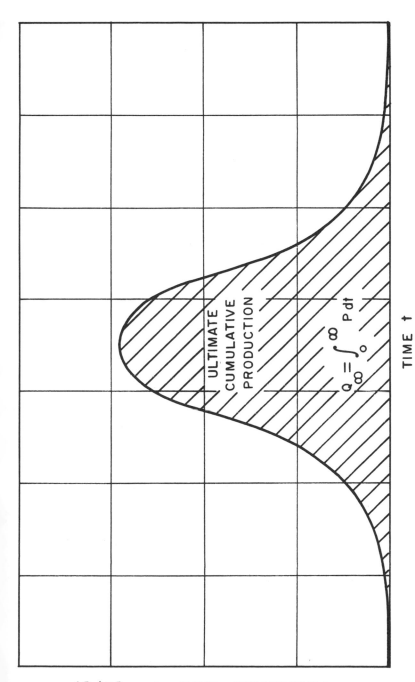

$$Q_\infty = \int_0^\infty P\,dt$$

ULTIMATE CUMULATIVE PRODUCTION

TIME t

PRODUCTION RATE: P = dQ/dt

Fig. 2.8 Curve of complete cycle of production of an exhaustible resource.
SOURCE: Hubbert 1956, *Drilling and Production Practice (1956)*, fig. 11, p. 12.
By permission of the American Petroleum Institute.

ties of minable coal in the major geographical regions of the world are those shown graphically in figure 2.9 that were published in 1969 by Paul Averitt of the United States Geological Survey.[1] These figures represent 50 percent of the initial coal in place in beds of more than 0.35 meters in thickness, and to depths up to two kilometers.

According to these estimates, the initial world supply of minable coal amounted to 7.64×10^{12} metric tons, of which 65 percent occurs in Asia (including European USSR), 27 percent in North America, 5 percent in western Europe, and only 2.4 percent in the three entire continents of Africa, South America, and Australia. Of this initial quantity, 135×10^9 metric tons, or 1.8 percent, had been mined by the end of 1969.

Using the foregoing estimate of the world's initial minable coal, three different extensions of the production rate into the future are shown in figure 2.10. The first of these is a continuation of the production curve at the same growth rate of 3.56 percent per year that has prevailed since 1945. The second two curves are based on the principle illustrated in figure 2.8, the first assuming that the ultimate cumulative production, Q_∞, will equal the 7.6×10^{12} metric tons of the Averitt estimate, and the second a lower figure of 4.3×10^{12}. Of these two figures for the assumed ultimate production, the smaller one may be the more realistic in view of the thinness and depth of some of the coal seams included in the Averitt estimates.

The complete cycle of coal production can be divided into three significant time periods: the very long times required to produce the first and the last ten percentiles of the ultimate quantity, Q_∞, and the much shorter period required to produce the middle 80 percent. As the curves in figure 2.10 indicate, with only a modest future increase in the rate of production, the period required to consume the middle 80 percent will probably not be longer than the three or four centuries between the years 2000 and 2300 or 2400.

United States crude oil

Like coal, the petroleum group of fossil fuels is derived from the unoxidized remains of former plant and animal life, and hence accumulations of these fuels are found either within or immediately adjacent to basins of sedimentary rocks. The petroleum group of fuels, by chemical composition, are composed of hydrocarbons of varying degrees of complexity. Physically, they include: natural gases, principally methane (CH_4); low-density natural-gas liquids produced with natural gas; crude oil, which is the liquid produced by oil wells; heavy viscous oils or tars; and petroleum solids such as kerogen in oil shales and vein deposits such as gilsonite.

Up to the present, the principal petroleum fuels produced have been crude oil, natural gas, and natural-gas liquids. Unlike coal, petroleum fluids

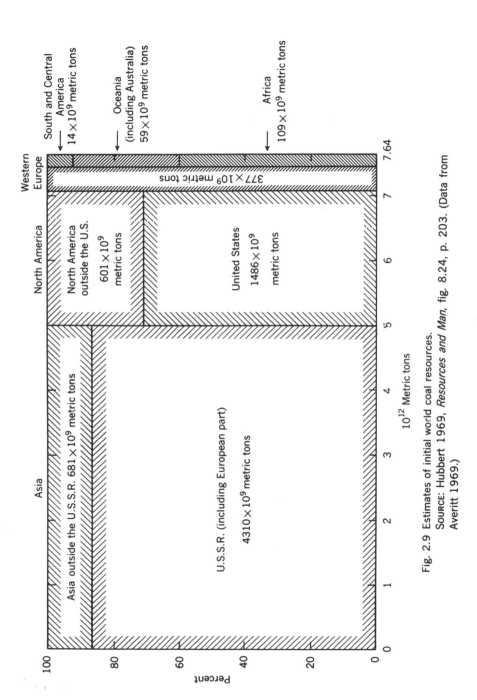

Fig. 2.9 Estimates of initial world coal resources. SOURCE: Hubbert 1969, *Resources and Man*, fig. 8.24, p. 203. (Data from Averitt 1969.)

23

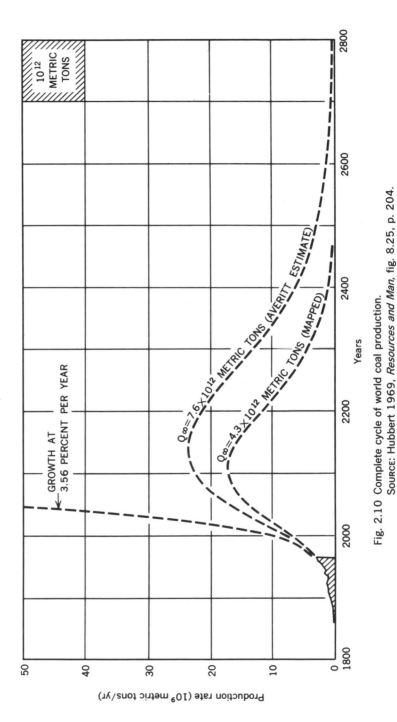

Fig. 2.10 Complete cycle of world coal production.
Source: Hubbert 1969, *Resources and Man*, fig. 8.25, p. 204.

24

are mobile and accumulate underground in restricted volumes of three-dimensional space in response to the combined influences of the forces exerted by the surrounding ground water and the channels and barriers to migration within the sedimentary rocks. For this reason, the estimation of the amount of oil or gas that a given region may ultimately produce is a much more hazardous undertaking than the estimation of coal.

Methods of estimation of the ultimate amount of oil or gas that a given region may produce reduce essentially to two stages: (1) the estimation for primary areas which depends principally upon the empirical results of exploration and production in those areas, and (2) the estimation of secondary areas based initially upon geological and geophysical mapping and geological comparisons with primary areas, and later by the cumulative results of exploratory and production activities.

As a primary area, the United States, exclusive of Alaska, is the world's leading petroleum-producing country and is also the most intensively explored major petroleum-producing region of the world. For this reason, estimates of the ultimate amount of oil that will be produced in the United States are commonly used as a partial basis in making estimates for the other petroleum-producing regions. Consequently, published world estimates have tended to be roughly proportional to those made by the same authors for the United States.

For estimating the amount of oil, Q_∞, that the United States may ultimately produce, a number of procedures are available that give substantially the same results. One of the better of these consists of relating cumulative discoveries to the cumulative depth of exploratory drillings.[2] In this case, the amount of oil discovered during a given year represents the ultimate amount of oil the fields discovered in that year will produce. If Q represents the ultimate amount of oil to be produced by fields discovered by the cumulative depth h of exploratory drilling, then dQ/dh will represent the amount of oil found per unit depth of incremental drilling.

Hence, instead of plotting a curve of dQ/dt versus cumulative time t, we may plot a curve of dQ/dh versus cumulative depth of exploratory drilling h. Each of these two curves has the same mathematical property that the area under the curve up to time t, or cumulative drilling h, respectively, represents the cumulative discoveries to that point. Of these two types of curves, however, the second has the advantage that it is principally dependent upon the technology of exploration and production, and is relatively insensitive to economic influences.

By 1966, the cumulative exploratory drilling in the United States amounted to 15×10^8 ft. In figure 2.11, the average rates of oil discovery per foot for successive 10^8-ft. intervals of drilling are plotted as abscissas versus cumulative drilling as ordinates. From the time scale along the top

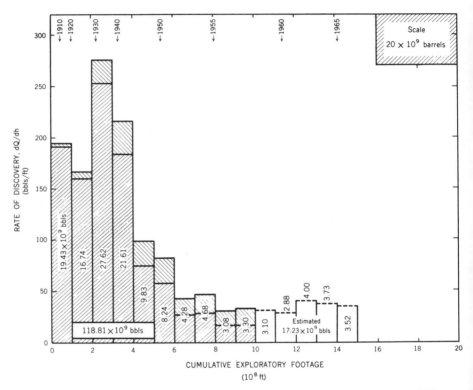

Fig. 2.11 Average United States discoveries of crude oil per foot of exploratory drilling for each 10^8-ft. interval of exploratory drilling.
SOURCE: Hubbert 1967, *AAPG Bulletin* 51, no. 11, fig. 15, p. 2223.

of the graph, it will be seen that the rate of drilling has been highly nonlinear with respect to time. The first 10^8-ft. interval, for example, required the entire sixty-year period from 1860 to 1920; the last few intervals have averaged about two years each.

What we see from figure 2.11 is that during the first three 10^8-ft. drilling intervals extending from 1860 until 1938, oil discovery in the United States was comparatively easy. During the first interval, with shallow wells, an average discovery rate of 194 bbls./ft. was achieved. During the second interval, from about 1920 to 1929, this dropped somewhat to 167 bbls./ft. Then, during the third interval from about 1929 to 1938, which included the first wide-spread use of geophysical methods of exploration and of electrical well logging as well as the fortuitous discovery of the large East Texas field, it reached a maximum rate of 276 bbls./ft. Following that, the discovery rate has declined drastically to a present figure of about 35 bbls./ft., notwithstanding the fact that during the period since 1938 the United States petroleum industry has engaged in the most intensive program of research and development of exploration and production techniques in its entire history.

By 1966, the cumulative discoveries, defined as the estimated ultimate production of all known fields, amounted to 136×10^9 bbls. A conservative extrapolation of the decline curve of figure 2.11, based on a negative-exponential curve passing through the ordinates of the first and last drilling intervals, gives an estimate for Q_∞ of about 165×10^9 bbls.

Using the figure of 165×10^9 bbls., we can now construct a curve of United States crude-oil production based on the principle of figure 2.8. This is shown in figure 2.12. In this, the dashed curve at the top shows what the production would have been had it continued to increase at the average exponential rate of 5.86 percent per year that had prevailed from about 1933 to 1955. The vertical dashed line at 1934 marks the date at which the cumulative production reached 16.5×10^9 bbls., the first 10 percent of the estimated value of Q_∞. That at 1999 marks the date at which Q_d is estimated to reach 90 percent of Q_∞. It is significant, therefore, that the time span required to produce the middle 80 percent of the oil resources of the United States will be only about sixty-five years—less than a human lifetime.

United States natural gas

Because natural gas and crude oil are genetically related, the amount of natural gas still to be discovered in the conterminous United States can be estimated from our previous estimate of the crude oil still to be discovered, in conjunction with the national average of the gas-to-oil ratio. At present, the ratio of the gas discovered during a period of a few years to the crude oil discovered during the same period amounts to about 6,500 ft.^{3}/bbl. This

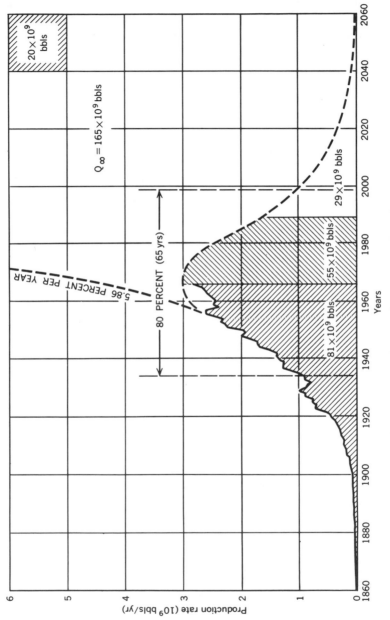

Fig. 2.12 Complete cycle of United States crude-oil production.
Source: Hubbert 1969, *Resources and Man*, fig. 8.17, p. 183.

28

ratio tends to increase with deeper drilling, which also is increasing. Making a liberal allowance for this increase, we shall here assume a ratio of 7,500 ft.^{3}/bbl. for future discoveries.

By the end of 1969, the data on natural-gas discoveries in the conterminous United States were the following:

Cumulative production, Q_p	370×10^{12} ft^3
Proved reserves, Q_r	275×10^{12} ft^3
Cumulative proved discoveries, Q_d	645×10^{12} ft^3.

If we now assume that the additional amount of gas contained in fields already discovered bears approximately the same ratio of 0.75 to the proved reserves as was found for crude oil, we obtain an estimate of 206×10^{12} ft.^{3} for this additional amount of gas. Adding this to the cumulative proved discoveries then gives 851×10^{12} ft.^{3} as our estimation for the amount of gas that will ultimately be produced from fields already discovered by the end of 1969.

By the end of 1969, the total amount of oil estimated to be producible from fields already discovered was 141×10^9 bbls. This represents 83 percent of the 165×10^9 bbls. ultimately to be produced, leaving but about 24×10^9 bbls. still to be discovered. For future discoveries, allowing 7,500 ft.^{3}/bbl. for the 24×10^9 barrels of oil still to be discovered, the figure of 180×10^{12} ft.^{3} for future gas discoveries is obtained. Adding this to the 851×10^{12} ft.^{3} already discovered then gives 1031×10^{12} ft.^{3} as our estimate for Q_∞, the ultimate amount of gas to be produced in the conterminous United States and its adjacent continental shelves.

In figure 2.13, the complete cycle of United States natural-gas production (exclusive of Alaska), based on an assumed value of 1290×10^{12} ft.^{3} is shown. This figure is the estimate made in 1967 by the Potential Gas Committee and is slightly higher than that given above.

United States natural-gas liquids

Natural-gas liquids are the low-density petroleum liquids produced as byproducts of natural-gas production. At present, in the United States 1 bbl. of natural-gas liquids is being produced with each 26,000 ft.^{3} of natural gas. Since this ratio is slowly increasing, we assume a higher value of 1 bbl./25,000 ft.^{3} for future production. Applying this to future gas production, we obtain the figure of 35×10^9 for the ultimate production of natural-gas liquids. Adding this to the figure of 165×10^9 bbls. for crude oil, we obtain an estimate of 200×10^9 bbls. for the ultimate production of petroleum liquids in the conterminous United States and its adjacent continental shelves. The full cycle of production of petroleum liquids based on this figure is shown in figure 2.14, which indicates that the peak of the smoothed production-rate curve should occur near 1970.

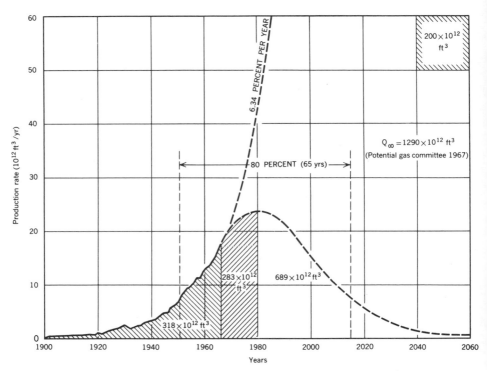

Fig. 2.13 Complete cycle of United States production of natural gas.
SOURCE: Hubbert 1969, *Resources and Man,* fig. 8.20, p. 190.

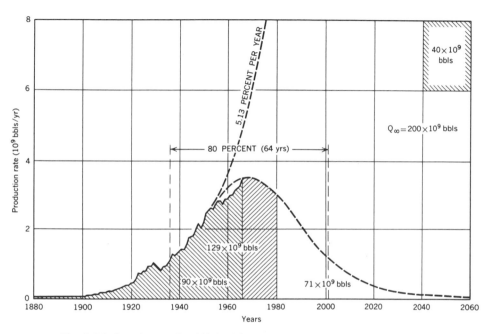

Fig. 2.14 Complete cycle of United States production of petroleum liquids.
SOURCE: Hubbert 1969, *Resources and Man,* fig. 8.22, p. 193.

Alaska

In Alaska, petroleum production in the Cook Inlet basin was begun in 1958. On the Alaska North Slope, the undeveloped Umiat field on the U.S. Naval Reserve No. 4 was discovered in 1947 and confirmed by additional drilling in 1950. The Prudhoe Bay field, with an estimated 10×10^9-bbl. oil reserve, was discovered in 1967. Since that time, there has been feverish exploratory activity on the North Slope. However, enough information on which to base a preliminary judgment of the magnitude of the Alaska petroleum resources has not been publicly released yet. Moreover, it probably will not be possible to make better than a speculative estimate until development has proceeded for a few more years.

From such information as is now available, it appears that the ultimate amount of crude oil that Alaska may produce will probably not be less than 30×10^9 bbls. Assuming this figure, the proportionate amounts of natural gas and natural-gas liquids would amount to 180×10^{12} ft.^{3} and 6×10^9 bbls., respectively.

While these quantities are significant, they still represent only a few years supply for the petroleum and natural-gas requirements of the United States.

World crude oil

As we have remarked before, the exploration and development in most other oil-bearing regions of the world are much less advanced than in the United States. Consequently, estimates of the ultimate amounts of oil that these areas may produce must be based jointly upon geological comparisons with the more advanced primary areas and the empirical production of the secondary areas themselves. In table 2.2 are two separate estimates of the ultimate crude-oil production of the major oil-bearing areas of the world obtained in 1967 from W. P. Ryman, Deputy Exploration Manager, Standard Oil Company of New Jersey. The first of these, totaling $2,000 \times 10^9$ bbls., was attributed to an unpublished report by L. G. Weeks; the second is a 1967 revision by Ryman of the Weeks estimate, totaling $2,090 \times 10^9$ bbls.

It is here considered that the Ryman 1967 estimate is probably as good an estimate of the *relative* amounts of oil the various areas of the world may produce as can be compiled from present information. However, it follows closely the earlier estimates by Weeks which included a figure of 270×10^9 bbls., for the United States. Since this is about one and one-half times more than our present estimate for the United States, it is possible that the Weeks estimate for the rest of the world may be high by about the same factor. Therefore, in figure 2.15 giving the complete cycle of world crude-oil production, two figures for Q_∞ have been used, the Ryman figure rounded off

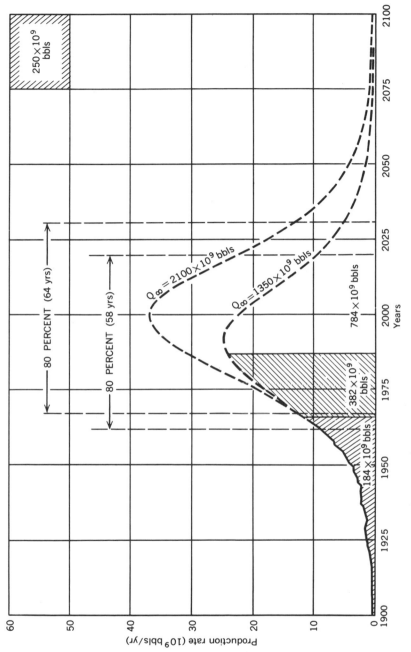

Fig. 2.15 Complete cycle of world crude oil production.
SOURCE: Hubbert 1969, *Resources and Man*, fig. 8.23, p. 196.

33

TABLE 2.2
Estimates of Ultimate Amounts of Crude Oil To Be Produced
in Major World Regions

Country or Region	Estimates by	
	L. G. Weeks, 1962 (10^9 bbls.)	W. P. Ryman, 1967 (10^9 bbls.)
United States	270	200
Canada	85	95
Latin America	221	225 *
Western Europe	19	20
Africa	100	250
Middle East	780	600
Far East	85	200 *
USSR, China, and satellites	440	500
World totals	2,000	2,090

SOURCE: W. P. Ryman, Deputy Exploration Manager, Standard Oil Company of New Jersey in M. King Hubbert, "Energy Resources," *Resources and Man* (San Francisco: W. H. Freeman and Co., 1969), pp. 157–242.
* Includes additional offshore areas.

to 2,100 $\times$ 10^9 bbls., and a lower figure of 1,350 $\times$ 10^9 bbls., which is about two-thirds of the former. The uncertainty of present estimates is considered to be roughly within those limits. In figure 2.15, it is seen that for the larger figure for Q_∞, the peak of world production would probably occur at about the year 2000, and the middle 80 percent would be produced within the sixty-four-year period between 1968 and 2032. For the smaller figure, the production peak would probably occur about 1990, and the middle 80 percent be produced during the fifty-eight-year period between about 1961 and 2019.

World natural gas and natural-gas liquids

Since the quantities of natural gas and of natural-gas liquids for large productive areas are roughly proportional to those of crude oil, the ultimate quantities of these fuels may be based on the estimates of crude oil. Such estimates for the low figure of 1,350 $\times$ 10^9 bbls. for crude oil and for the high figure of 2,100 $\times$ 10^9 bbls. are given in table 2.3. The quantities of natural gas corresponding to these two figures are about 8,000 $\times$ 10^{12} ft.^{3} and 12,000 $\times$ 10^{12} ft.^{3}, respectively. Corresponding estimates for natural-gas liquids are 250 $\times$ 10^9 and 420 $\times$ 10^9 bbls. Adding the natural-gas liquids to the crude oil estimates gives estimates of 1,620 $\times$ 10^9 bbls. and 2,520 $\times$ 10^9 bbls. for the low and high estimates of total petroleum liquids.

TABLE 2.3
Estimates of Ultimate World Production of Fossil Fuels

	Crude Oil (10^9 bbls.)	Natural-Gas Liquids (10^9 bbls.)	Total Petro-leum Liquids (10^9 bbls.)	Natural Gas (10^{12} ft.^{3})
Low estimate	1,350	250	1,620	8,000
High estimate	2,100	420	2,520	12,000

SOURCE: Hubbert, "Energy Resources," *Resources and Man.*

Tar sands and oil shales

So-called tar, or heavy-oil, sands are sands containing oil that is too viscous to permit recovery by natural flowage into wells. Since such sands are as yet mostly unexploited, no world inventory of their occurrences is available. However, the best known of such occurrences, and possibly the world's largest, are those in the northern part of the Province of Alberta, Canada. These consist of the large Athabasca deposit and two smaller groups, the Bluesky-Gething and the Grand Rapids. Recent estimates of the evaluated reserves of these three areas are given in table 2.4. These include approximately 267×10^9 bbls. for the Athabasca deposit and 34×10^9 for the two smaller groups combined, or a total of 301×10^9 bbls. of potentially producible oil.

The first large-scale mining and extraction operation of the Athabasca deposit was begun by a group of oil companies in 1966. Doubtless others will follow as soon as crude-oil production begins to fall short of requirements.

Oil shales differ from tar sands in that their contents of hydrocarbons consist of the solid kerogen rather than of viscous liquids. This solid, however, distills out as a vapor upon heating and then condenses to a liquid.

TABLE 2.4
Tar-Sand Deposits of Alberta, Canada

Area	Estimated Resources (10^9 bbls.)
Athabasca	266.9
Bluesky-Gething	20.6
Grand Rapids	13.3
Total	300.8

SOURCE: J. R. Pow, G. H. Fairbanks, and W. J. Zamora, "Description and Reserve Estimates of the Oil Sands of Alberta," in *Athabasca Oil Sands,* Research Council of Alberta, Information Series no. 45 (1963), pp. 1–14.

The best known and possibly the largest of such deposits are those of the Green River oil shales occurring in separate basins in western Colorado, southwestern Wyoming, and eastern Utah. The oil contents of various strata of these shales range from near zero for the poorest to as high as 65 U.S. gallons per short ton (0.25 m³/met. ton) for the richest. According to Duncan and Swanson,[3] the total oil content of these shales in the medium range of 10–25 gal./ton amounts to about 1,430 × 10⁹ bbls. Of this, however, they list only 80 × 10⁹ bbls. as being "recoverable under present conditions."

A world inventory of known oil shales by the same authors is given in table 2.5. According to these estimates, the total oil contents of shales

TABLE 2.5
Known Shale-Oil Resources of World Land Areas

Grade (U.S. gals./ton)	Recoverable under 1965 Conditions	Marginal and Submarginal		
	10–100	5–10	10–25	25–100
Continents	Oil Content (10⁹ barrels)			
Africa	10	small	small	90
Asia	20	ne	14	70
Australia and New Zealand	small	ne	1	small
Europe	30	ne	6	40
North America	80	2,200	1,600	520
South America	50	ne	750	small
Totals	190	2,200	2,400	720

SOURCE: Duncan and Swanson, "Organic-rich Shales," United States Geological Survey.
ne: No estimate.

containing between 10 and 100 U.S. gallons of oil per short ton (0.04 to 0.4 m³/met. ton) amount to about 3,120 × 10⁹ bbls. Of this, however, they regard only 190 × 10¹² bbls. (including the 80 × 10⁹ bbls. for the Green River shale) to be recoverable under present conditions.

SUMMARY OF THE FOSSIL FUELS

The foregoing conclusions pertaining to the initial world supply of the various fossil fuels capable of being produced under present conditions of technology are summarized in table 2.6. Here, in addition to the estimated magnitudes of the various fuels expressed in units of mass or volume, the energy contents are also given, expressed in both thermal joules and thermal

TABLE 2.6
Energy Contents of the World's Initial Supply of Recoverable Fossil Fuels

Fuel	Quantity	Energy Content		Percentage
		(10^{21} thermal joules)	(10^{15} thermal kilowatt-hours)	
Coal and lignite	7.6×10^{12} metric tons	201	55.9	88.8
Petroleum liquids	$2,000 \times 10^9$ bbls. (272×10^9 metric tons)	11.7	3.25	5.2
Natural gas	$10,000 \times 10^{12}$ ft.^{3} (283×10^{12} m^3)	10.6	2.94	4.7
Tar-sand oil	300×10^9 bbls. (41×10^9 metric tons)	1.8	0.51	0.8
Shale oil	190×10^9 bbls. (26×10^9 metric tons)	1.2	0.32	0.5
Totals		226.3	62.9	100.0

kilowatt-hours. It is significant that, of the total energy 226 $\times$ 10^{21} thermal joules of all the fossil fuels, 201 $\times$ 10^{21}, or 89 percent, is represented by coal and lignite, and only 25 $\times$ 10^{21}, or 11 percent, by the entire petroleum group of fuels. It is also of interest that the energy contents of petroleum liquids and of natural gas are approximately equal: about 10 $\times$ 10^{21} thermal joules, or 5 percent each.

As we noted earlier, of the initial supply of 7.6 $\times$ 10^{12} metric tons of coal, only 135 $\times$ 10^9 metric tons, or 1.8 percent, had been consumed by the end of 1969. For crude oil, the cumulative world consumption by the end of 1969 amounted to approximately 227 $\times$ 10^9 bbls. If we assume a figure intermediate between our previous low and high estimates of 1,670 $\times$ 10^9 bbls. for the initial quantity of crude oil, then the oil already consumed amounts to 13.6 percent of this initial quantity.

One of the most important results of this evaluation of the world's supply of fossil fuels is the contrast that it affords between the geological time of some 700 million years required for their accumulation, and the brief three or four centuries of human history required for their annihilation.

To appreciate the bearing of this on the long-range outlook for human institutions, the historical epoch of the exploitation of the world's supply of fossil fuels is shown graphically in figure 2.16, plotted on a time scale extending from 5,000 years ago to 5,000 years in the future—a period well within the prospective span of human history. On such a time scale, it is seen that the epoch of the fossil fuels can only be a transitory and ephemeral event—an event, nonetheless, which has exercised the most drastic influence experienced by the human species during its entire biological history.

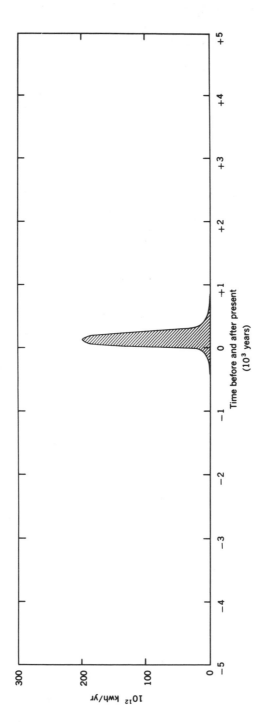

Fig. 2.16 Fossil fuels in human history.
Source: Hubbert 1969, *Resources and Man*, fig. 8.27, p. 206.

38

OTHER CONVENTIONAL SOURCES OF ENERGY

For other sources of energy appropriate for power production either in parallel with, or as successors to, the fossil fuels, we again direct our attention to the terrestrial flux of energy depicted in figure 2.1. The thermal power influx from the solar radiation of $178,000 \times 10^{12}$ watts dwarfs the inputs of 32×10^{12} and 3×10^{12} watts from the other two principal sources, geothermal energy and tidal energy, respectively. It also is approximately 31,000 times the world's 1969 rate of production of industrial energy of 5.7×10^{12} watts.

Solar power

Despite the total magnitude of solar power, and the fact that it is responsible for most of the material circulations on the earth, its low areal density makes the direct use of solar power impractical and prohibitive in cost for other than small-scale, special-purpose uses. This can be illustrated by considering the building of a solar-electric power plant of 1,000 electrical megawatts capacity. With a conversion factor from solar to electrical power of 0.1, such a plant would require a solar-power input of 10^{10} watts. According to Daniels, the solar energy incident upon the earth's surface amounts to about 500 (cal./cm.^{2})/day.[4] This, when averaged over a full day, gives an average solar input of 2.4×10^{-2} watts/cm.^{2}. The area of the earth's surface required to collect 10^{10} watts of solar power would accordingly be 42 km.^{2}, or the area of a square of 6.5 km. per side.

There is no question that it would be physically possible to cover such an area with energy-collecting devices, and to transmit, store, and ultimately transform the energy so collected into conventional electric power. However the complexity of such a process, and its costs in terms of metals and of chemical and electrical equipment, as well as of maintenance, in comparison with those for thermoelectric and hydroelectric plants of the same capacity, render such an undertaking to be of questionable practicability.

Therefore, the large-scale uses of solar energy appear to be limited principally to the biological channel of photosynthesis and to the natural mechanical concentrations of wind and water power. Of these, wind power is also impractical for any but small-scale special purposes, such as windmills and the propulsion of sailing vessels, and even these uses have largely been supplanted by other sources of power.*

---

* Since the foregoing passage was written there has been a great deal of interest on the part of scientists, engineers, and research groups in the possibility of the development of large-scale solar power.

Water power

For large-scale power production, therefore, we are left only with that fraction of solar power that is concentrated by the circulation of water during the hydrologic cycle. Since the potential water power of any given site is proportional jointly to the magnitude of the stream discharge and to the height of fall, the world's best water-power sites are concentrated in regions of heavy rainfall and of large topographic relief.

In table 2.7, a compilation is given of the potential water-power capacities of the various continents or major geographical and political areas of the world. Also shown are the approximate amounts of developed water-power capacities of these same areas for the year 1967, compiled from a recent Federal Power Commission report on *World Power Data, 1967.*[5]

According to these data, the world's total potential water-power capacity is about 3×10^{12} watts, which is of the same order of magnitude as the world's present rate of industrial-energy consumption. However, only 8.5 percent is developed at present, and this principally in the highly industrialized areas of North America, Western Europe, and Japan. The two continents with the largest potential water-power capacities are Africa and South America with capacities of 780×10^3 and 577×10^3 megawatts, respectively. These capacities are also the least developed of all of those of the world's major areas. Moreover, to develop and utilize such sources of power requires a simultaneous industrialization of the respective regions, and this, in turn, requires quantities of industrial metals roughly proportional to the

TABLE 2.7
World Potential and Developed Water-Power Capacity

Region	Potential Power[a] (10^3 Mw.)	Percentage of Total	Developed[b] Capacity, 1967 (10^3 Mw.)	Percentage Developed
North America	313	11	76	23
South America	577	20	10	1.7
Western Europe	158	6	90	57
Africa	780	27	5	0.6
Middle East	21	1	1	4.8
Southeast Asia	455	16	6	1.3
Far East	42	1	20	48
Australia	45	2	5	11
USSR, China, and satellites	466	16	30	6.4
World	2,857	100	243	8.5

SOURCES: a. Francis L. Adams, United States Federal Power Commission, 1961.
b. United States Federal Power Commission, *World Power Data, 1967.*

power developed. Whether the earth's resources of metallic ores will be adequate for such an expansion remains to be seen.

Viewed superficially, water power has the appearance of being an essentially inexhaustible source of energy, or at least one with a time scale comparable to that required to remove mountain ranges by erosion. Actually, this may not be so. Most water-power projects involve the damming of streams and the construction of large reservoirs. The time required for most of these reservoirs to become filled with sediment will only be from one to a few centuries. Hence, unless a satisfactory solution to this problem should eventually be found, most of the world's water-power capacity may prove to be comparatively short-lived.

Tidal power

Tidal power is similar to water power except that it is derived from the alternate filling and emptying, with the semi-diurnal period of 12 hours 24.4 minutes of the synodical lunar day, of a bay or estuary that can be enclosed by a dam. When a tidal basin is enclosed, the maximum power obtainable would be by a flow cycle that permitted the basin to fill and to empty during brief periods at high and low tide. For such a cycle, the energy potentially obtainable per cycle would be given by

$$E_{\max} = \rho g \, R^2 S,$$

where ρ is the water density, g the acceleration of gravity, R the tidal range, and S the area of the enclosed basin.

The average power for a complete cycle would be, therefore,

$$\bar{P} = E_{\max}/T = \rho g \, R^2 S/T,$$

where $T = 4.46 \times 10^4$ sec. is the tidal period, which is also the half period of the synodical lunar day. Of this, the theoretical maximum amount realizable in engineering calculations has commonly been taken to be from about 8 to 20 percent. However, in the French La Rance project the power realized is reported to approach 25 percent of this maximum amount.

In table 2.8, a summary is given of the average potential tidal power and the annual amount of energy that could be obtained from the world's most favorable tidal-power sites. These include individual sites with potential power capacities in the range of from 2 to 20,000 megawatts each. The total potential power capacity of all of these sites is 64,000 megawatts. This represents 2 percent of the total tidal energy dissipation of 3×10^{12} watts. It is also only 2 percent of the world's potential water-power capacity.

The world's first major tidal-power installation is that of the La Rance estuary in France, which began operation in 1966. This had an initial capacity of 240 Mw. which is planned to be increased to a total of 320 Mw.

TABLE 2.8
Tidal-Power Sites and Maximum Potential Power

Locality or Region	Average Potential Power (10^3 kw.)	Potential Annual Energy Production (10^6 kwh.)
North America/Bay of Fundy/(9 sites)	29,027	255,020
South America/Argentina/San José	5,870	51,500
Europe/England/Severn	1,680	14,700
France (9 sites)	11,149	97,811
USSR (4 sites)	16,049	140,452
Totals	63,775	559,483

SOURCES: a. N. W. Trenholm, "Canada's Wasting Asset—Tidal Power," *Electrical Engineering News* 70 (1961): 52–55.
b. Lev B. Bernshtein, *Tidal Energy for Electric Power* [English trans. of 1961 Russian ed.] Israel Program for Science Translations (Jerusalem, 1965).

Geothermal power

Unlike tidal power and water power which depend upon continuing sources of energy, geothermal power depends essentially upon the "mining" and eventual depletion of temporarily stored quantities of volcanic heat. Such sources of thermal energy are accordingly associated with the volcanic regions of the world.

A summary of the developed and planned geothermal-electric power installations of the world as of 1969 is given in table 2.9. At that time, the total installed capacity amounted to 828 Mw., with planned additional capacity that would raise this to 1,342 Mw. by the early 1970s. Most of the geothermal-electric plants are comparatively small—in the range of 1 to 20 megawatts. The three localities with larger installations are Larderello, Italy; The Geysers in California; and Wairakei, New Zealand. The development of geothermal power at Larderello was begun in 1904, and has reached a present capacity of 370 Mw. Power production at The Geysers was begun in 1960 with a small 12.5 Mw. unit. Capacity has subsequently been increased to 82 Mw. by 1969. According to the 1969 Annual Report of the Magma Power Company, which operates this project, two additional 55 Mw. plants are under construction and are scheduled for completion by mid-1971. This would increase present capacity to 192 Mw. Plans are also under way to increase the total capacity to 400 Mw. by 1973. The plant at Wairakei, New Zealand, began operation in 1958 and has a present capacity of 290 Mw., which is reported to be about the maximum limit for this locality.

As to the world's ultimate geothermal capacity, after extensive studies of the approximate magnitudes of the world's major geothermal areas,

TABLE 2.9
Developed and Planned Geothermal-Electric Power Installations

Country and Locality	Installed Capacity 1969 (megawatts)	Planned Additional Capacity (megawatts)	Total Capacity by Early 1970s (megawatts)	Date of Earliest Installation
Italy				
Larderello	370		370	1904
Monte Amiata	19		19	ca. 1962
Total	389		389	
United States				
The Geysers,				
California	82	318	400	1960
New Zealand				
Wairakei	290		290	Nov. 1958
Mexico				
Pathé	3.5		3.5	ca. 1958
Cerro Prieto, Mexicali		75	75	ca. 1971
Total	3.5	75	78.5	
Japan				
Matsukawa	20	40	60	Oct. 1966
Otake	13	47	60	Aug. 1967
Goshogate		10	10	
Total	33	97	130	
Iceland	(Geothermal energy			
Hveragerdi	for house and			
	greenhouse heating)	17	17	1960
USSR/Kamchatka				
Pauzhetsk	5	7.5	12.5	1966
Paratunka	0.75		0.75	1968
Bolshiye Bannyye	25		25	1968
Total	30.75	7.5	38.25	
Grand total	828.25	514.5	1,342.75	

SOURCE: Hubbert, "Energy Resources," *Resources and Man* (1969).

Donald E. White of the United States Geological Survey has estimated that the stored thermal energy in such areas to depths of 10 km. amounts to about 4×10^{20} thermal joules.[6] With a 0.25 conversion factor this would yield 1×10^{20} joules of electrical energy, or about 3×10^6 Mwe.-yrs. Then, if this amount of energy were to be withdrawn during a period of fifty years, the average electric power produced would be 60,000 Mw., which is about sixty times the present installed geothermal-electric capacity.

It thus appears that the maximum magnitude of geothermal-electric power capacity will probably be in the range of tens of thousands of electrical megawatts, but that this source of energy will probably be largely depleted in less than a century. While 60,000 Mw. is a significant quantity of power, a better idea of its significance can best be obtained by comparison with other sources. It is approximately equal to the world's potential tidal power, but only about 2 percent of the world's potential water power. Also we may compare the initial quantity of 4×10^{20} joules of geothermal energy with that of the fossil fuels as given in table 2.6. This represents less than 2 percent of the initial energy of the fossil fuels, and only 20 percent of that of petroleum liquids and natural gas combined.

NUCLEAR POWER

The one major source of energy of a magnitude appropriate for large-scale power production whose resources remain to be investigated is that of nuclear energy—the energy obtainable by the fissioning of certain isotopes at the upper end of the scale of atomic masses, and by the fusion of others at the lower end.

Fission

During the incredibly short period of twenty-eight years since the first controlled fission chain reaction was achieved on 2 December 1942, the development of nuclear reactors, based principally on the fissioning of uranium-235, has progressed to the point where, by the end of 1969, fifteen central-station nuclear power plants were in operation in the United States. These have individual capacities ranging from 22 to 575 electrical mega-watts and a combined capacity of 3,482 Mwe. At the same time, according to the United States Atomic Energy Commission, eighty-two additional plants with individual capacities mostly within the range of 500 to 1,000 Mwe., and an aggregate capacity of 70,000 Mwe., were either under construction or contract.[7] By 1980, the AEC estimates that the nuclear-power capacity of the United States will reach 150,000 Mwe., which will be approximately 25 percent of the total electric-power capacity of the country.

An increase in ten years from a power capacity of 3,482 to 150,000 Mwe. represents an exponential growth rate of 37.6 percent per year with a doubling period of only 1.84 years.

In a joint report of January 1969 the European Nuclear Energy Agency and the International Atomic Energy Agency show that the central-station nuclear power capacity of the entire non-Communist world has also been increasing with a doubling period of about two years, and had reached a total capacity of 20,000 megawatts by the end of 1969.[8] It was assumed that after 1972 the growth rate will slow down, and that by 1980 the total capacity will be within the range of 220,000 to 340,000 Mwe.

Uranium and thorium requirements and resources

Estimations of the uranium requirements to meet this anticipated growth in nuclear-power capacity have been made by the United States Atomic Energy Commission and jointly by the European Nuclear Energy Agency and the International Atomic Energy Agency. According to Rafford L. Faulkner, Director of the Division of Raw Materials, AEC, the United States cumulative requirements of U_3O_8 will reach 212,000 short tons by 1980 and 450,000 by 1985.[9] For the non-Communist world outside the United States, he estimates that the cumulative requirements for the period 1969–1980 will be 212,000 short tons and that this will increase to 490,000 by 1985. The ENEA–IAEA in their joint report of 1969, estimate that the cumulative requirements of the non-Communist world by 1980 will be between 563,000 and 739,000 short tons of U_3O_8.

These are the requirements against which the resources of uranium must be weighed. According to a recent AEC news release, the United States reserves of U_3O_8 minable at less than \$8/lb. amounted to 204,000 short tons at the end of 1969. During 1969, as the result of 29.9×10^6 feet of drilling, 56,000 tons of U_3O_8 were discovered. This was at a discovery rate of 3.7 lbs./ft., which is less than half the average rate of 8 lbs./ft. that had prevailed up until a few years previously.

For the reserves of U_3O_8 of the non-Communist countries, producible at \$10/lb., the ENEA-OAEA estimated a total of 700,000 short tons as of 1969. In a review published in 1970 Robert D. Nininger of the AEC gave the estimates of the reserves of seventeen different countries which are summarized in table 2.10.[10] His total estimate was 980,000 short tons. Of

TABLE 2.10
Non-Communist World 1970 Resources of U_3O_8 Producible at Less Than \$10/lb.

Countries	Reserves of U_3O_8 (tons)	Percentage of Total
United States	250,000	25.5
Canada	200,000	20.4
France and territories	188,000	19.2
South Africa	205,000	20.9
Subtotal	843,000	86.0
13 other countries	137,000	14.0
Grand total	980,000	100.0

SOURCE: Nininger, "The World Uranium Picture" (Address before Colorado Mining Association and 73rd National Western Mining Conference), U.S. Atomic Energy Commission (1970).

this, 843,000, or 86 percent, were accounted for by four countries, the United States, Canada, France and territories, and the Union of South Africa. The remaining 14 percent were distributed among thirteen other countries.

From such estimates it appears that the uranium resources now known are adequate for the requirements of the non-Communist world for approximately the next fifteen years. Present productive capacity, however, is not sufficient to meet the annual requirements beyond the year 1975. Although new deposits of uranium will undoubtedly continue to be discovered, the fact remains that uranium is a comparatively rare element, and its deposits producible in the low-cost range must occur in limited quantities. Hence, the uranium requirements of a nuclear-power capacity that doubles every two years can rapidly deplete all likely deposits of this nature.

For this reason, were the nuclear power reactors of the future to be of the same types as those at present, which consume only one or two percent of natural uranium, it is doubtful whether the episode of nuclear power could last for much longer than a century. It is imperative, therefore, if this consequence is to be avoided, that the present type of burner or low-conversion reactors based on the rare isotope uranium-235, be supplanted by breeder reactors capable of consuming completely the fertile isotopes, uranium-238 and thorium-233—that is, the whole of natural uranium and thorium. Should this be accomplished, then not only would the producible energy from the present supplies of high-grade uranium be enhanced by a factor of fifty to 100, but that of the much larger supplies of lower-grade deposits of both uranium and thorium would become available. This chapter does not attempt a review of these low-grade deposits, though such a review has been made by McKelvey and Duncan.[11] The following two examples, however, will convey some idea of the order of magnitudes of the energy involved.

In the United States, the Chattanooga shale of Devonian age crops out along the western border of the Appalachian Mountains in eastern Tennessee and neighboring states, and underlies much of the areas of a half-dozen midwestern states at comparatively shallow depths. In its outcrop area in eastern Tennessee, according to Vernon E. Swanson of the United States Geological Survey, this shale has a uranium-rich stratum, the Gassaway member, which is about five meters thick and contains about 60 grams of uranium per metric ton.[12] Even in Oklahoma, about 1,000 kilometers to the west, the stratigraphic equivalent of this shale is also uranium rich as determined by gamma-ray logging in oil wells.

In the Tennessee area, the uranium content of this shale would amount to 150 grams per cubic meter or to 750 grams per square meter of horizontal surface for the five-meter thick stratum. Assuming the use of breeder reactors, the energy content per gram of natural uranium would be 8.2×10^{10}

joules, or an amount equivalent to the energy content of 2.7 metric tons of bituminous coal, or to 13.7 barrels of crude oil. Therefore, the energy content of this shale per square meter of horizontal surface would be equivalent to that of about 2,000 metric tons of coal, or 10,000 barrels of crude oil. The energy content of an area thirteen kilometers square would be equivalent to that of the world's resources of crude oil, and that of an area sixty kilometers square would be equivalent to that of the world's coal resources.

For a thorium resource, the Conway granite in New Hampshire, which crops out over an area of 750 km.[2] and probably extends to depths of several kilometers, contains 150 grams of thorium per cubic meter.[13] This is equivalent to 400 tons of coal or 2,000 barrels of oil per cubic meter. The energy content of a surface layer of this rock 100 meters thick would be equivalent to that of 100 times the world's crude-oil resources, or to four times that of the world's coal.

The world's supply of uranium and thorium in rocks containing fifty grams or more per metric ton is probably tens of times greater than that in the two examples just cited. Consequently, with the exclusive use of breeder reactors, the world's supply of energy appropriate for power production would probably be of the order of tens to hundreds of times greater than that of the fossil fuels.

Fusion

Since controlled fusion has not yet been achieved, only a brief mention of the energy from this source will be made. At present, one of the more promising fusion reactions is that of deuterium and lithium-6. Omitting intermediate details, the end result of this reaction is the following:

$$^6_3\text{Li} + {}^2_1\text{D} \rightarrow 2\,{}^4_2\text{He} + 22.4 \text{ Mev.}$$

The use of this reaction will be limited either by deuterium or lithium-6, whichever may be the scarcer. The natural abundance of deuterium in sea water is approximately one atom of deuterium for each 6,700 atoms of hydrogen. From this it may be computed that one cubic meter of sea water contains 10^{25} atoms of deuterium, and the entire oceans, 1.5×10^{43}. A large fraction of this could be extracted by methods now available at an energy cost of but a small fraction of that releasable by fusion.

Lithium deposits, on the other hand, occur on land. Lithium is produced from the geologically rare igneous rocks known as pegmatites and from the salts of saline lakes. The known lithium resources of North America and

Africa were recently reviewed with James J. Norton, the specialist of the United States Geological Survey on this subject. According to Norton, these contain a total of about 1.6×10^6 metric tons of Li_2O. Thomas L. Kesler, a geologist for the Foote Mineral Company, the largest producer of lithium in the United States, has published estimates of about 2×10^6 metric tons for the United States, 390,000 for Canada, and 180,000 for Africa.[14,15] Allowing for future developments, Norton considers 10×10^6 metric tons of elemental lithium to be a good order-of-magnitude figure for the presently known resources of lithium for these areas.

Lithium-6 represents only 7.42 atom percent of natural lithium, and one metric ton of lithium-6 contains 1.0×10^{29} atoms.

The resources of Li_2O in the United States, Canada, and Africa, based on the data of Norton and Kesler, are given in table 2.11. This includes a total of 19.6×10^6 metric tons of Li_2O of which the lithium-6 content is 675×10^3 metric tons which contain 6.75×10^{34} atoms of lithium-6. It appears, therefore, that even if we limit the extractable deuterium in sea water to 10 percent of the total content, the quantity of lithium-6 available is only about 10^{-8} of that of deuterium. Therefore, the energy of 22.4 Mev of the deuterium–lithium-6 fusion reaction may be entirely ascribed to the lithium-6. On this basis, the fusion of the 6.75×10^{34} atoms of lithium-6 in table 2.11 will produce 2.4×10^{23} joules of thermal energy. This is about the same as the total thermal energy of the fossil fuels, as given in table 2.6.

TABLE 2.11
Estimated Lithium Reserves

Location	Li_2O Measured, Indicated, and Inferred (10^6 metric tons)	Lithium Metal (10^6 metric tons)	Lithium-6 (10^4 metric tons)	Number of Lithium-6 Atoms (10^{33} atoms)	Equivalent Fusion Energy (10^{21} joules)
United States	19.0	8.8	65.4	65.4	234
Canada	0.4	0.2	1.4	1.4	5
Africa	0.2	0.1	0.7	0.7	2.5
Total	19.6	9.1	67.5	67.5	241.5

Sources: a. James J. Norton, personal communication.
  b. Kesler, "Lithium Raw Materials," in *American Institute Mining Metallurgical Petroleum Engineering.*
  c. Kesler, "Exploration of King's Mountain Pegmatites," *Mining Engineering.*

TRANSITIONAL PERIOD OF INDUSTRIAL EXPONENTIAL GROWTH

From this review of the world's energy resources appropriate for power production, two realizations of outstanding significance emerge. One is the brevity of the time, as compared with the totality of past human history, during which the large-scale production of power has grown; the other is that energy resources sufficient to sustain power production of present magnitudes for at least a few millenia have now become available.

The limiting factors in power production are, therefore, no longer the scarcity of energy resources, but rather the principles of ecology. The production of power and its associated industrial activities are quite as much components of the world's ecological complex as are the populations of plant and animal species—witness, for example, the decline of the population of horses with the rise of motor vehicles. As we have observed, our industrial activities have been characterized by large exponential rates of growth only for the last two centuries, and the present rate of growth of nuclear power capacity, with a doubling period of but two years, represents the most spectacular growth phenomenon in the entire history of technology. However, it is equally as true with power capacity and automobiles as with biologic populations that the earth itself cannot sustain or tolerate any physical growth for more than a few tens of doublings. Therefore, any individual activity must either cease its growth and, as in the case of water power, stabilize at some maximum or intermediate level that can be sustained, or else it must pass a maximum and decline eventually to extinction.

Because of the impossibility of sustaining rates of industrial growth such as those which have prevailed during the last century and a half, it is inevitable that before very much longer such growth must cease, and some kind of stability be achieved. As indicated in figure 2.17, the future period of stability could be characterized either by a continuation of a technological culture with a high level of energy consumption, or by a cultural decline to a primitive low-energy level of existence.

Regardless of which of these possible courses may actually be followed, it is clear that the episode of industrial exponential growth can only be a transitory epoch of about three centuries duration in the totality of human history. It represents but a brief transitional epoch between two much longer periods, each characterized by rates of change so slow as to be regarded essentially as a period of nongrowth. Although the forthcoming period poses no insuperable physical or biological difficulties, it can hardly fail to force a major revision of those aspects of our current economic and social thinking which stem from the assumption that the growth rates that

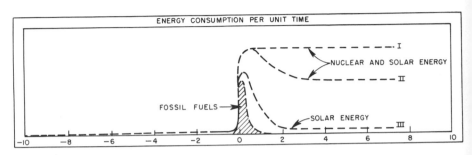

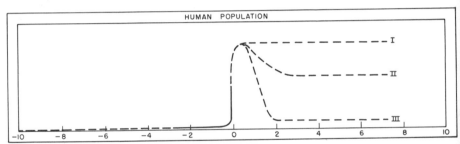

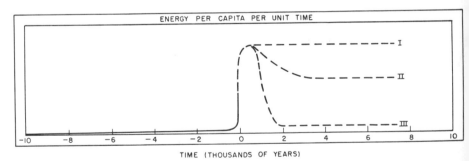

TIME (THOUSANDS OF YEARS)

Fig. 2.17 Epoch of industrial growth in context of longer span of human history.
SOURCE: Hubbert 1962, *Energy Resources*, fig. 61, p. 134.

have characterized this temporary period somehow can be made permanent.

Reference notes

1. Paul Averitt, "Coal Resources of the United States, January 1, 1967," *United States Geological Survey Bulletin* 1275 (Washington, D.C., 1969): 116.
2. M. King Hubbert, "Degree of Advancement of Petroleum Exploration in United States," *American Association Petroleum Geologists Bulletin* 51, no. 11 (1967): 2207–27.
3. D. C. Duncan and V. E. Swanson, "Organic-rich Shales of the United States and World Land Areas," *United States Geological Survey Circular* 523 (Washington, D.C., 1965).
4. Farrington Daniels, *Direct Use of the Sun's Energy* (New Haven and London: Yale University Press, 1964).
5. Federal Power Commission, *World Power Data, 1967* (Washington, D.C., February 1969).
6. Donald E. White, "Geothermal Energy," *United States Geological Survey Circular* 519 (Washington, D.C., 1965).
7. United States Atomic Energy Commission, *Annual Report to Congress for 1969* (Washington, D.C., 1970).
8. European Nuclear Energy Agency and International Atomic Energy Agency, *Uranium Production and Short Term Demand* (Paris and Vienna, January 1969).
9. Rafford L. Faulkner, "Uranium Supply and Demand" (Address before American Mining Congress, San Francisco, 19 October 1969), United States Atomic Energy Commission, Washington, D.C.
10. Robert D. Nininger, "The World Uranium Picture" (Address before Colorado Mining Association and 73rd National Western Mining Conference and Exhibition, Denver, Colo., 13 February 1970), United States Atomic Energy Commission, Washington, D.C.
11. V. E. McKelvey and D. C. Duncan, "United States and World Resources of Energy," in *Symposium on Fuel and Energy Economics, Joint with Division on Chemical Marketing and Economics,* 149th National Meeting American Chemical Society, Division Fuel Chemistry, vol. 9, no. 2 (1965) pp. 1–17.
12. V. E. Swanson, "Oil Yield and Uranium Content of Black Shales," *United States Geological Survey Professional Paper* 356A (Washington, D.C., 1960).
13. J. A. S. Adams, et al., "The Conway Granite of New Hampshire as a Major Low-grade Thorium Resource," in *Proceedings of the National Academy of Sciences,* vol. 48 (1962), pp. 1898–1905.
14. Thomas L. Kesler, "Lithium Raw Materials," in *American Institute of Mining, Metallurgical and Petroleum Engineers,* ed., Joseph L. Gillson (New York, 1960), chap. 24, pp. 521–31.
15. Thomas L. Kesler, "Exploration of the Kings Mountain Pegmatites," *Mining Engineering* 13, no. 9 (1961): 1062–68.

3. Air Resources

A. J. Haagen-Smit
Chairman, Air Resources Board, State of California

Air pollution has been a common companion of man's activities. Recent instances of more serious pollution episodes, as occurred in London and Donora, may be singled out for special mention. Today, however, it is not necessary to look far for examples of the effects of men on air resources; all metropolitan areas are afflicted with an alarming deterioration of air quality.

Air and water are present in such abundance that we rarely appreciate them until the supply diminishes or the addition of various gases and other substances make these media unpalatable. There have been warnings, however, that we should guard the air as a precious resource. A handbook on nutrition written around the turn of the century devoted a whole chapter to the importance of air as a part of man's food. Dr. König, the author of the handbook, writes of the quantity of air processed through our lungs per day, which is more than all of our solid and liquid foods combined. The average person inhales about half a liter with every breath; repeating this sixteen times per minute, or 23,000 times per day, this amounts to about twenty-five pounds. It is not surprising that Dr. König discusses the impurities in the air and their effects on human well-being just as he had done for other food stuffs. This prophetic deviation from accepted procedures— considering air as a part of man's food—was undoubtedly prompted by the author's experiences in his hometown, Münster, a center of industrial activities in Westphalia, Germany.

LOS ANGELES SMOG

Some twenty years ago I was reminded of these writings when I decided to investigate the causes of the chemical composition of the now world-famous Los Angeles smog. I had just finished a research project determining the volatile flavor components of pineapple fruit by passing the vapors through a set of cold traps and analyzing the condensate. This setup was ideally suited for the analysis of the smog. The only thing I had to do was to leave out the fruit and pass the outside air through the traps. After processing about 400 cubic feet of air, I had collected in the traps a few ounces of blackish liquid, mostly water, of course, but water which had the characteristic flavor of smog in highly concentrated form. Anyone would refuse to drink this water; nevertheless, that glass of water is what we process through our lungs every day. Since lungs are extremely efficient in adsorbing miscellaneous materials, Dr. König's warning that we should guard the quality of the air as we do that of food is most pertinent.

There is no doubt that we can repeat this experiment in any of our metropolitan areas. The composition of the collected material will be different, but the general conclusions are the same.

Why do we pollute the air on a nationwide and global scale today? The main reason, of course, is the explosive growth of the population and our failure to adjust our way-of-living and our technology to this fact. Every nine seconds there is a birth; every sixteen and one-half seconds someone dies. This means our population grows by four every minute. Every year there are nearly two million more people in the United States. At the same time the estimated four billion in the world have increased by more than fifty million.

If each world citizen received an equal share of the land, he would possess about ten acres of real estate. This should keep him in food for quite a long time, even though he soon will have to divide his land among exponentially increasing numbers of offspring. But what about the air? There are about 200 billion times a billion cubic feet of air. This seems like a large amount but we share it with about four billion other people, and the personal share for each person on earth would be about 60 billion cubic feet. This is sufficient for some time to come, but the problem is that there are some people that do not have access to all that air at all times.

Geographic conditions such as living in a valley or bowl surrounded by mountains may severely decrease the general exchange of air. When this is coupled with a meteorological condition, usually called *inversion,* the available air may be severely limited. This inversion condition is characterized by having a cold layer of air at ground level which does not mix with the higher levels; this is a frequent situation. These so-called *temperature inversions occur in all parts of the world,* not only in widely-advertised Los Angeles. When such conditions prevail, the air available to people is contained as though in a flat box, the height of which may vary from a few feet to several thousand feet.

Under such an inversion condition, the capacity of the Los Angeles area to disperse its pollutants is greatly limited. For example, if an inversion height is only a thousand feet, there would be only five million cubic feet of air available for each of the ten million inhabitants of Los Angeles instead of the sixty billion cubic feet available to the world citizen. This seems like sufficient air until we realize that for breathing alone we consume or make unfit for use, 400 cubic feet of air daily and about ten times as much, or 4000 cubic feet in miscellaneous operations, such as driving an automobile, cooking, heating, and manufacturing products. Air conditioning experts tell us that it would be advisable to have about 1,000 times as much air as that which is being used up, which means that we require at least five million cubic feet of air to cover our daily needs. This is only the quantity in the Basin available to the Los Angeles inhabitant, and unless the air is continuously renewed, the conclusion is inevitable that the air quality is destined to deteriorate to unacceptable levels. Complaints of citizens, as well as the results of the pollution monitoring system, give abundant proof that two-thirds of the year the ventilation in the Los Angeles area is not enough to live up to the minimum recommended levels of air quality. The limited capacity of the air to accommodate our effluents is accentuated in our canyon-like city streets. Carbon monoxide, for example, produced in automobile traffic, establishes concentrations of ten to thirty parts per million (V/V). On heavily traveled streets, concentrations of 100 parts per million are quite normal (fig. 3.1). Similarly, concentrations of other auto emissions are raised and the presence of large quantities of unburned gasoline, oils, and products of incomplete combustion in city air is readily shown. Superimposed on this blanket of traffic pollutants are the effluents of the combustion of fuel in industry and homes and the emissions of thousands of miscellaneous industries.

Estimated nationwide emissions are listed in table 3.1.[1] This table does not list the major contribution to the air from human activities, carbon dioxide. The burning of fossil fuels adds two billion tons yearly to the atmosphere. In some industrial areas the carbon dioxide content of the air increased from a normal 300 ppm. to as high as 700 ppm. Fortunately, photosynthetic activities of plants partially offset this increase. Nevertheless, a steady increase of carbon dioxide has been observed, which together with particulate matter, may affect the heat balance of the earth.

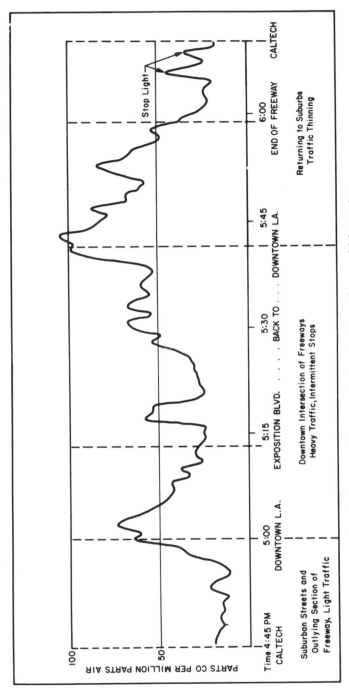

Fig. 3.1 Carbon monoxide concentration in city and freeway driving routes.
Source: A. J. Haagen-Smit, "Carbon Monoxide Levels in City Driving," *Archives of Environmental Health* 12 (1966): 548–51.

TABLE 3.1
Estimated Nationwide Emissions, 1968
(in millions of tons per year)

Source	Carbon Mon-oxide	Partic-ulates	Sulfur Oxides	Hydro-carbon	Nitro-gen Oxides	Total	Percent-age of Total
Transportation	63.8	1.2	0.8	16.6	8.1	90.5	42.3
Fuel combustion in stationary sources (power generation, industry, space heating)	1.9	8.9	24.0	0.7	10.0	45.5	21.3
Industrial processes	9.7	7.5	7.3	4.6	0.2	29.3	13.7
Solid waste disposal	7.8	1.1	0.1	1.6	0.6	11.2	5.2
Miscellaneous (agri-cultural burning, forest fires)	16.9	9.6	0.6	8.5	1.7	37.3	17.4
Total tons	100.1	28.3	32.8	32.0	20.6	213.8	100.

SOURCE: *Nationwide Inventory of Air Pollution Emissions—1968,* National Air Pollution Control Administration Publication AP-73 (August 1970).

SMOKE POLLUTION

One of the most common forms of air pollution is that caused by smoke. A cross-country trip by airplane should convince anyone that smoke is a problem in every large city from Los Angeles to New York. Large streamers, extending for scores of miles, have their origin in open-burning dumps, lumberyards, steel mills, foundries, and power plants, to name only a few of the multitude of sources. This is an old problem. Some 200 years ago laws were passed in England to regulate the burning of certain types of coal. Even today combustion is still a major source of air pollution.

At first sight the problem of controlling the emission of smoke appears quite simple, but we have only to think how tobacco smoke bubbles, seemingly undisturbed, through water to realize that this problem is not so easily solved. Large particles of soot and dust can readily be removed in settling chambers, or in cyclones, where gravitational or centrifugal forces are used to separate the particles by weight. Other methods consist of filtering out the dust through cloth made of cotton, woven plastics, or even glass fibers, in structures known as baghouses. For very fine dust, in the order of from 0.5 to 1.0 micron, we have to apply processes whereby the electrically-charged particle is removed by electrostatic forces by passage between electrodes kept at a potential difference from 15,000 to 40,000 volts. A comparison of particle size of effluents and the means of their recovery is shown in figure 3.2.

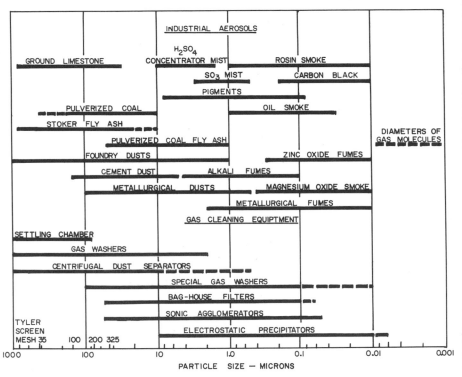

Fig. 3.2 Relative size of well-known particulates and the effective range of various control methods.

SOURCE: H. P. Munger, "The Spectrum of Particle Size and Its Relation to Air Pollution," *Air Pollution* 2, chap. 16.

By catching the particles of a size close to the wavelength of light, an important fraction of the dust responsible for the scattering of light is removed, and therefore, the probability of visible plume formation is substantially decreased. Each one of these dust-collection processes has found wide application in industry, but even today systematic studies to increase their efficiency would be most welcome.

The ash left after combustion of the fuel is not the only agent responsible for objectionable emissions from smokestacks. Virtually all other constituents of flue gas—water, carbon monoxide, and oxides of sulfur and nitrogen—play a role. In the atmosphere they are joined by contributions from chemical factories, automobiles, and miscellaneous burning operations. The nature of these gases varies greatly. The main pollutant gases are the oxides of sulfur and nitrogen, hydrocarbons, and carbon monoxides from the incomplete combustion of gasoline. In addition, there are hundreds of minor organic components originating in various combustion processes. The more reactive gases can be removed by wet processes using water scrubbers or chemical sprays. Dry processes pass the polluted air over beds or through towers filled with reactive chemicals such as lime. In special cases when cost is subordinate and a high removal efficiency must be obtained, charcoal absorbers are indicated. The sometimes unpredictable outcome of the application of these systems is evidence that a more thorough basic investigation in the operation of the collecting equipment is needed. This is especially pressing in view of the strict health standards and the increase in size of operations which require considerably higher efficiency and higher capacity at an economically acceptable cost.

WASTE PROBLEM

One of the problems which immediately becomes obvious is what can be done with the material once it is collected? The pilot plant may collect a few hundred cubic feet per minute but nature operates with hundreds of thousands of cubic feet a minute. Whatever the final product, we must dispose of it, perhaps in the form of sulfur, ammonium sulfate, sulfuric acid or calcium sulfate. The market is limited for these products, especially when the conversions are carried out on a national scale.

The total sulfur dioxide released in this country from power plants and smelters is in the order of thirty-two million tons a year—a procession of twenty-ton trucks which would reach a few times around the earth. Where do we leave the collected material? The waste problem is faced by everyone in the air, water and soil field. We must learn that the production process is not finished with producing the desired product be it energy, plastics, or eggs. *Cleaning up and eventual recycling must become a part of the price we pay for our standard-of-living. It needs the best brains we have.*

ATMOSPHERIC REACTION

Smoke and dust problems remain the cause of most pollution complaints, but bothersome new problems have arisen. Chemicals released into the air are subject to atmospheric influences: sunlight, oxygen, and the action of other pollutants. These problems became rather acute in the case of Los Angeles smog, but a similar problem has been studied earlier in the atmospheric reactions of a totally different kind which had been noticed in studying the well-known bluish plumes of uncontrolled power plants burning sulfur-containing fuel. The sulfur dioxide leaving the stacks is oxidized to the trioxide which reacts with water to form sulfuric acid. The reactions do not stop there, however; calcium carbonate and ammonia neutralize the acid and the final result of the SO_2 release is the presence of calcium and ammonia sulfate, contributing to the haziness which is often noticeable at considerable distances from industrial establishments.

It was clear that the eye-irritating smog present in Los Angeles must have a different origin; this became abundantly clear when the efforts in reducing dustfall and the emission of the sulfur oxides did not give relief from the smog symptoms. The analysis of the condensate I had collected during the heavy-smog day pointed to its origin from organic materials.

The isolation of these organic oxidation products, such as aldehydes and organic acids, led to the hypothesis that Los Angeles smog was formed through the oxidation of gasoline and other organics released in large quantities in the Los Angeles Basin. Smog is usually accompanied by haze and a strong oxidizing action manifested by its effect on rubber. Rubber goods were known to be subject to heavy cracking in the Los Angeles area and this action was similar to that caused by ozone. A convenient way of measuring this oxidizing effect made use of potassium iodide as a reagent; drawing smog air through a solution of the iodide liberates iodine, which can be titrated or measured colorimetrically or electrically.

Figure 3.3 shows the daily recurrence of the high oxidant during three successive days which were followed by two days when rain cleared the skies.

Other effects of smog were extensive damage to the leaves of plants, quite different from that caused by the well-known effects of pollutants such as sulfur dioxide or fluorine. Finally, there was the frequent complaint of eye irritation aggravated by a simultaneously appearing bluish haze. The easiest way to test the hypothesis that smog is formed in an atmospheric oxidation by hydrocarbons was to combine gasoline with a gaseous oxidation agent,

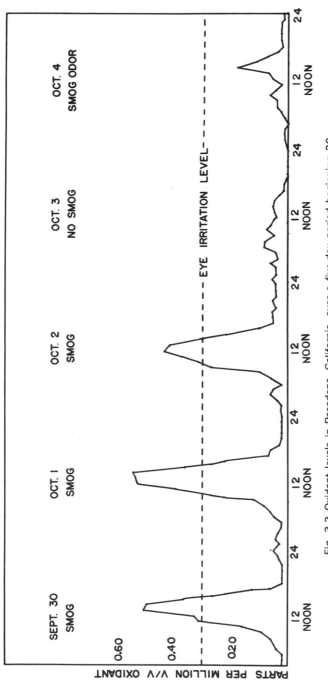

Fig. 3.3 Oxidant levels in Pasadena, California, over a five-day period beginning 30 September 1953.

SOURCE: A. J. Haagen-Smit, "Chemistry and Physiology of Los Angeles Smog," *Industrial and Engineering Chemistry* 44 (1952): 1342–46.

such as strongly oxidizing ozone, and to blow this mixture in a fumigation room where people and plants could be exposed. The results were most gratifying: the subjects had to wipe their eyes, the plants developed the typical symptoms, and odor and haze added to the impression that we had created some real smog. There was, however, one bothersome problem: where did the ozone come from?

OZONE FORMATION

During smog attacks the ozone content of the Los Angeles air reaches levels twenty to thirty times greater than found in nonurban areas. Concentrations of half a part of ozone per million parts of air have repeatedly been measured during heavy smogs. To establish such a concentration directly would require the dispersal of more than a thousand tons of ozone in the Los Angeles Basin. No industry released significant amounts of ozone; discharges from electric power lines were also negligible, amounting to less than a ton a day. A considerable amount of ozone is formed in the upper atmosphere by the action of short ultraviolet rays, but that ozone does not descend to earth during smog conditions because of the very temperature inversion that intensifies smog and which forms a barrier not only to the rise of pollutants but also to the descent of ozone.

Exclusion of these possibilities leaves sunlight as the only suspect in the creation of the Los Angeles ozone. The cause cannot be direct formation of ozone by sunlight at the earth's surface as that requires radiation of wavelengths shorter than 2,000 angstrom units, which do not penetrate the atmosphere to ground level. There was a compelling reason, however, to look for an indirect connection between smog and the action of sunlight; high oxidant or ozone values are found only during daylight hours. Apparently a photochemical reaction was taking place when one or more ingredients of smog were exposed to sunlight—which is, of course, abundant in the Los Angeles area.

In order for a substance to be affected by light, it has to absorb the light, and the energy of the light quanta has to be sufficiently high to rupture the chemical bonds of the substance. It is not likely that this substance would be a simple hydrocarbon. The energy required to split the C-C bond or the C-H bond would be in the order of 84 and 100 kilocalories per mole. These energies are available in ultraviolet light, but daylight would not suffice to remove an electron from these bonds.

A more likely candidate for an atmospheric photochemical reaction in smog is nitrogen dioxide. This dioxide is formed from nitrogen oxide, which originates in all high temperature combustion through a combining of the nitrogen and oxygen of the air. Nitrogen dioxide has a brownish color and absorbs light in the region of the spectrum from the blue to the near ultraviolet. Radiation from the sun can readily dissociate nitrogen dioxide into nitric oxide and atomic oxygen. The reactive oxygen formed in the photochemical dissociation, according to equation a, can attack organic material and could readily be responsible for the formation of peroxydic products, aldehydes, and acids found in our smog condensates. And true enough, replacing ozone in our earlier fumigation experiments with nitrogen dioxide and carrying out the reactions in bright sunlight resulted in a reproduction of all the smog symptoms. Of special importance was the observation that bent pieces of rubber suspended in the fumigation room showed deep cracks identical to those caused by the action of ozone, confirming my guess that ozone was formed by the formation of free radicals through the action of atomic oxygen on hydrocarbons (eq. b). These radicals surrounded by oxygen molecules would form peroxyl radicals and these in turn with a second molecule of oxygen could form ozone (eq. c and d). These reactions are shown in table 3.2.

TABLE 3.2
Schematic Summary of the Main Mechanism Leading To
Photochemical Smog Symptoms

(a) $NO_2 + h\nu \rightarrow$	$NO + O$	photodissociation of nitrogen dioxide
(b) $O + RH \rightarrow$	$HO\cdot + R\cdot$	alkyl radical formation
(c) $R\cdot + O_2 \rightarrow$	$ROO\cdot$	peroxyl radical formation
(d) $ROO\cdot + O_2 \rightarrow$	$RO\cdot + O_3$	ozone formation
(e) $ROO\cdot + NO_2 \rightarrow$	$ROONO_2$	peracyl nitrate formation
(f) $O + O_2 \rightarrow$	O_3	ozone formation

The essential part of the ozone-forming reaction, formation of organic radicals, can be readily demonstrated by irradiating a substance known to dissociate in radicals upon irradiation with sunlight: *diacetyl.*

$$\underset{\substack{\| \quad \| \\ O \quad O}}{\overset{\text{Diacetyl}}{CH_3-C-C-CH_3}} \xrightarrow[\text{light}]{\text{sun}} \underset{\substack{\| \\ O}}{\overset{\cdot}{CH_3-C}} + O_2$$

$$\underset{\substack{\| \\ O}}{CH_3-C-O\cdot} + O_3 \xleftarrow[\text{ozone}]{+O_2} \underset{\substack{\| \\ O}}{CH_3-C-O-O\cdot}$$

I was quite familiar with this compound because of its overpowering butter aroma and also because it had been used for many years in various photochemical experiments. In all these experiments oxygen had always been excluded and the ozone-forming capacity had escaped attention. In the presence of air or oxygen, ozone is readily formed and can be detected spectrophotometrically or through its action on bent rubber strips. There is a quantitative relationship between the depth of cracking and the amount of ozone coming in contact with the rubber; this method, using rubber as a quantitative indicator of ozone concentration, has been useful not only in the theoretical investigations but in popularizing the unfamiliar smog reactions. It is possible during an hour of lecture to demonstrate with sensitive rubber the severe cracks caused by photochemically-produced ozone from diacetyl. The rubber-cracking test is still the least expensive method to demonstrate quickly if there is a photochemical smog problem in an area. Many theories have since been proposed to account for the formation of ozone in these reactions; they all have as an essential feature the primary reaction of photochemical dissociation of nitrogen dioxide and subsequent formation of organic radicals. The formation of radicals has received further support through the isolation of plant-damaging peracyl nitrates, combinations of peroxyl radicals and nitrogen dioxide (eq. e). Several elaborate schemes of formation of ozone and smog products have been published.[2]

FORMATION OF SMOG SYMPTOMS

A schematic view of the main reactions leading to the smog symptoms is shown in figure 3.4.

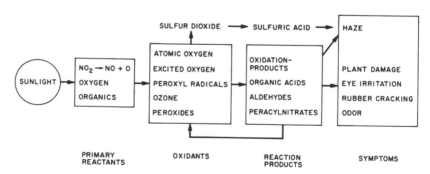

Fig. 3.4 Schematic presentation of smog formation.

Eye irritation, plant damage, ozone formation, and aerosol formation are all produced by the photochemical oxidation of the organic material. These reactions are not limited to hydrocarbons alone; derivatives, such as aldehydes and acids, contribute their share to the photochemical reaction products. The oxidant pool consisting of atomic oxygen, excited oxygen, peroxyl radicals, and ozone reacts also with sulfur dioxide, explaining its rather rapid oxidation with formation of sulfuric acid aerosols in polluted air.

MONITORING OF POLLUTANTS

The knowledge of the nature and quantities of the pollutants and the sources of their emissions is essential in the conservation of acceptable air quality. It is the task of the control agency to establish this information and develop the necessary control regulations.

One of the tools of control is measurement of the concentration of the pollutants in the field, as well as at the source of the emission. The rapidly developing interest in pollution control has stimulated the perfection of old methods and the invention of new methods for most of the pollutants by continuously-operating instruments. New methods for source testing have evolved for dealing with concentrations mostly in the order of a few hundred to a few thousand parts per million, but even more progress has been made in the analysis of the far lower concentrations of pollutants in ambient air. These concentrations measure in fractions of parts per million. For example, the concentration of sulfur dioxide in this area is in the order of 0.02 to 0.5 parts per million. Today in many areas of the country analytical instruments placed at strategic localities record the concentrations of the most important pollutants on a continuous basis. Records are made of sulfur dioxide, hydrocarbons, ozone, oxides of nitrogen, and particulate matter. A striking example of these new tools is the correlation spectrograph which can record various specific pollutants from a distance of several thousand feet. The peaks in figure 3.5, drawn with a correlation spectroscope, clearly show the contribution from the power plant complex at Long Beach and from automobiles at the civic center.

Testing has identified most of the emission sources and inventories of the emissions from the major polluters are now available for most of the larger cities. Emission factors of various operations are now well known and preliminary estimates can readily be made without testing. Combustion of fossil fuels is still a major source of smoke and oxides of sulfur, but in some areas the effluents of chemical factories, smelters, mineral processing plants, and open burning take the lead. A great deal of progress in the control of these sources has been made; however, in recent years the automobile has taken over the rank of major polluter in most urban areas.

AUTOMOBILE EMISSIONS

The emissions of the internal combustion engine are mainly due to incomplete combustion and escape of the unburned hydrocarbons and carbon monoxide through the exhaust and crankcase vent. The carburetor and tank add to the emission of hydrocarbons through evaporation of the fuel. In

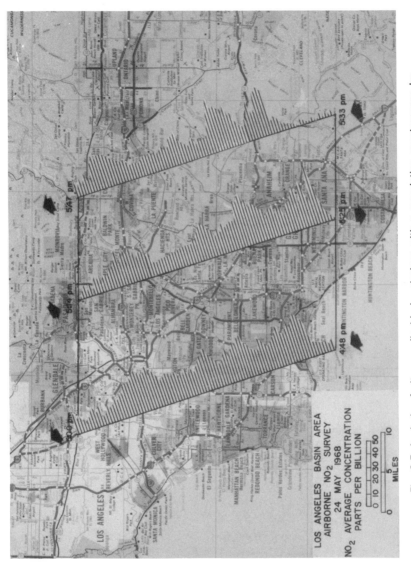

Fig. 3.5 Analysis of nitrogen dioxide by airplane with a correlation spectroscope by Barringer Corporation.
Source: A. R. Barringer, "Chemical Analysis by Remote Sensing," 23rd Annual ISA Instrumentation Automation Conference (October 1968).

addition, the high temperature of the explosion leads to relatively high concentrations of one of the primary factors in the smog reaction: the oxides of nitrogen. The relative importance of the emission sources for hydrocarbons, oxides of nitrogen, and carbon monoxide are shown in table 3.3.

TABLE 3.3
Relative Importance of Automobile Emission Sources

	HC	NO_x	CO
Exhaust	65%	100%	100%
Crankcase	20%	—	—
Evaporation losses, tank and carburetor	15%	—	—
Totals	100%	100%	100%

The total loss of fuel for the average driver using two gallons of gasoline per day is about 0.4 lbs., and loss in *fuel value* amounts to 15 percent when the large emission of carbon monoxide is included. The tremendous waste of fuel because of inefficiency of combustion by the ninety million automobiles in the United States alone is twenty-five million gallons per day. In the control effort, the auto industry has adopted a system which gives a more complete combustion in the engine itself by using a leaner fuel-air mixture, coupled with carburetor adjustments and retarding of the spark timing. The emission from the crankcase is now recirculated through the air filter and burned together with the normal flow of gasoline. The combined effect of these systems will reduce the hydrocarbon emissions from new passenger cars by about 80 percent. This is encouraging, but it does not mean that all of our problems are solved and that we have only to wait until the nonequipped older cars are off the road to see clear skies again.

CONTROL OF OXIDES OF NITROGEN

There are still many technical problems to solve and one of the most difficult tasks is the control of the oxides of nitrogen. The oxides of nitrogen are formed in high temperature combustions from nitrogen and oxygen of the atmosphere. In a survey conducted in Los Angeles County, the concentration of the oxides of nitrogen were determined from various sources ranging from small burners to large power plants.[3] Motor vehicles are by far the largest contributor in urban areas such as Los Angeles, but power plants and miscellaneous, small, but numerous fuel burners are contributing materially to the nitrogen oxide load in the Basin. Table 3.4 shows the number of tons per day of emission of oxides of nitrogen in the South Coast Basin in 1968.

TABLE 3.4
Emission of Oxides of Nitrogen in the South
Coast Basin of California, 1968 (tons per day)

Motor vehicles	843
Diesel vehicles	130
Chemical, metallurgical	6
Petroleum industry	57
Domestic and commercial heaters	60
Aircraft	24
Power plants	300
Ships and railroads	9
Incineration	8
Agriculture	1
Total	1,438

SOURCE: State of California Air Resources Board, *Emission Inventories* (Sacramento, Calif., November 1969).

Several engineering laboratories are now searching for ways to reduce the nitrogen dioxide content of exhaust gases from gas, oil, and gasoline combustion through studies on the variable combustion conditions which can be obtained by boiler modification or changes in engine design. The principles used in these control methods are based on a lowering of the flame temperature. In power plants a 50 percent reduction in oxides of nitrogen has been accomplished by the introduction of secondary combustion. The main body of the fuel is burned with a deficiency in oxygen; additional air is subsequently introduced above the burner, completing the combustion of the fuel. In recent years additional reductions in NO formation have been attained by exhaust gas recirculation, burner and boiler configuration and, finally, by operation of the plant below peak performance. In this way concentrations of oxides of nitrogen have been reduced to as low as fifty parts per million, a reduction by a factor of ten compared to uncontrolled boilers. Unfortunately, the capacity of the plants also increases rapidly and it will not be long before the gains made today are neutralized. It is for this reason that many pollution-minded people prefer a nuclear plant over a fossil one—granted though the choice is a hard one to make.

The story of control of the oxides of nitrogen from automobile exhaust has been a most disappointing one for those who believe in the necessity for control of the primary reactant in smog formation. The automobile industry, faced with legal requirements for carbon monoxide and hydrocarbon control, decreased the organics all right, but in the process the oxides of nitrogen were raised about 50 percent and beyond, partially neutralizing the advantage that would have been obtained from hydrocarbon control alone. Why this happened is shown in figure 3.6. The formation of ozone and eye irritants in the photochemical reaction is dependent on the presence of the oxides of nitrogen. These effects increase, reach a maximum, and then finally disappear at higher concentrations.

The automobile industry used this opportunity to reduce complaints, but apparently did not realize that the atmosphere beyond the maximum would be objectionable because of both odor and health reasons. Unfortunately, the attempts of the industry to reduce the hydrocarbons and carbon monoxide from the exhaust led to a marked increase in the oxides of nitrogen. This is shown in figure 3.7 which plots the production of hydrocarbons, carbon monoxide, and oxides of nitrogen in relation to the air:fuel ratio. The air:fuel ratio is the ratio of the weight of intake air and the weight of the fuel. The theoretical ratio for complete combustion is a little less than 15 to 1. With better combustion and less hydrocarbon and carbon monoxide in the exhaust, more oxides of nitrogen are formed. Only at very lean mixtures do all three exhaust components go down, and it is in this area that many inventors look for the ultimate answer to control. At present these attempts result in poor drivability; however, distinct progress is being made. Enriching the fuel results in an excess of unburned fuel in the exhaust, but the oxides of nitrogen decrease.

In automobile engines, promising methods return a fraction of the exhaust air back to the air intake and retard spark timing. In so doing, the flame temperature is lowered and fewer oxides of nitrogen are formed. The reduction can be as high as 80 percent without seriously affecting drivability.

Investigations are under way to find out if modification of the composition of gasoline may have some effect. There are now joint projects of the petroleum industry and the car manufacturers under way to tailor the gasoline to low smog rather than to more speed.

SMOG OR NO SMOG

Now we come to the billion dollar question: smog or no smog? In answering this question it is good to distinguish between the short-range outlook, the next ten years, and a longer-range view beyond that time, to perhaps the turn of the century. For the next ten years radical innovations, such as electric cars, turbine cars and steam cars, will not have much significance. We must rely on control of what we now have. That is, reduction of

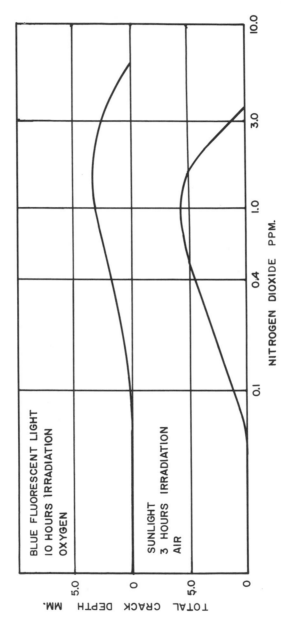

Fig. 3.6 Photochemical ozone formation with 3-methyl-heptane and varying concentrations of nitrogen dioxides hydrocarbon, concentration ten parts per million, ten hour exposure with blue fluorescent light.

70

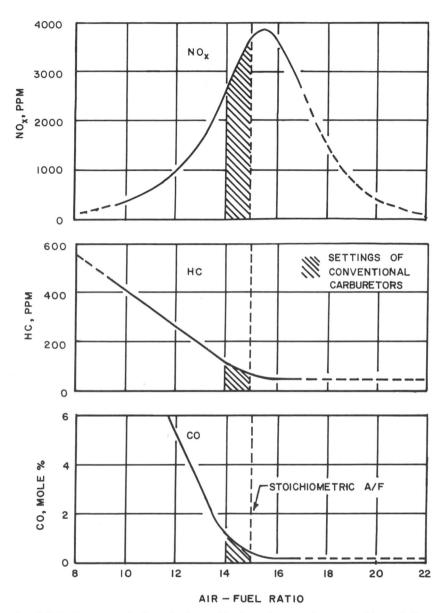

Fig. 3.7 Typical concentration of unburned hydrocarbons, carbon monoxide, and nitrogen oxides as a function of air-fuel ratio. Car speed 60 mph.

hydrocarbons and oxides of nitrogen in the present engines until we will have met the levels required by the health authorities. From an engineering point of view there is no reason why this cannot be done; there are indications that there is a genuine desire on the part of industry to solve this vexing problem.

But let us not forget that the car owner, too, has a responsibility. There is no excuse for smoking exhausts laying down smoke screens from one end of town to the other. The car owner has a responsibility for the performance of his own little power plant, just as the industry has for its operation. This is politically unappealing, however, educational work in this direction should continue.

Over the past seven years we have had a 60 to 70 percent control of exhaust hydrocarbons and nearly 100 percent control of crankcase gases. In 1970 California tightened the control of new cars by adopting evaporation-loss control from tank and carburetor, a 15 percent gain in hydrocarbon control. For the first time the truck owner was told that he, too, has to pitch in. The number of trucks is not as large as the numbers of automobiles, but their gasoline use and emissions are quite sizeable: about 10 percent of the total losses of all motor vehicles.

Slowly but surely, in a few years we will be near our goal of 80 to 90 percent reduction and we will have 75 percent control of carbon monoxide which will prevent trouble from this toxic agent in the foreseeable future. Disregarding the time it took, this is on the *plus* side, considering that we have not ruined the automobile industry and we still have acceptable automobiles.

Now let us look at the *negative* side. There are some irritating problems which detract from the theoretical efficiency; they concern the deterioration of control devices while the car is in use. A 50,000-mile accelerated test is not the same as 50,000 miles of private use. It is wishful thinking to expect that any system will be able to maintain its efficiency without professional correction. The manner in which this service can be organized is not clear at the moment. There are various possibilities, such as inspection stations, free or subsidized, where inspections are carried out by trained, competent personnel. The key, of course, is to have competent people who know how to inspect and to tune for low emissions, which is not the same as tuning by ear. This is a most important and, at the same time, a most troublesome problem.

It takes approximately ten years to remove the old cars from the roads, but it is discouraging to think that we have to wait so long before the auto problem is solved. Considering we do not have oxides of nitrogen control devices on the cars yet the job will not be finished until the middle of the 1980s. This date is not determined by Detroit; it is determined by the people. How much are they willing to pay for clean air and how much inconvenience are they willing to tolerate to adequately maintain their little smog machines? To shorten this waiting period we need other kinds of devices or methods that can take care of the used car market. Such devices may not be as effective as those put on new cars, but if they are inexpensive and easy to install, we must consider them. The same is true for possible gains that may be made by fuel alterations.

I have no doubt that the present county, state and federal laws are adequate to cope with the air pollution situation. I am also convinced that the engineering accomplishments to date will mean a satisfactory improvement in almost every part of our country. Unfortunately, I have to make an exception for my hometown and similar rapidly-growing metropolitan centers. The California South Coast Basin, of which Los Angeles is a part, will remain a problem child for many years to come, perhaps forever. It was there that eye irritation from photochemical smog was first noticed. Los Angeles has become an open air garage due to its sole reliance on the automobile. There are, however, other difficulties; the rapid growth of its population demands more power and more industry and causes more emissions by individuals from many activities, especially the heating and cooling of homes.

The federal government has set standards for ambient air which should be considered a goal for acceptable air quality. For the Los Angeles area those dealing with the oxides of nitrogen and the oxidant are of special interest. During days of heavy smog with an oxidant value higher than 0.4 ppm., the concentration of hydrocarbon and oxides of nitrogen are found to be between, respectively, 0.5–1.25 and 0.3–0.7 ppm. On days of no smog when the oxidant is lower than 0.1 ppm., the values of hydrocarbon and oxides of nitrogen are, respectively, 0.1–0.5 and 0.05–0.25 ppm. A series of such days are plotted in figure 3.8. These values are lined up so that the main axis of the ellipses point to the origin. The slight deviation is due to the inactive hydrocarbon background in the oil-rich county. The slope of the line is determined by measurements of the emission of the various sources. On a smog-free day, the dispersion of the pollutants is good and the concentration is very low; we are near the origin. On the other hand, when there is a low inversion, the concentrations are much higher. The ratio, however, between hydrocarbons and oxides of nitrogen remains constant. At an inversion height of a thousand feet, hydrocarbons and oxides of nitrogen establish concentration in the order of one part and 1/2 part per million, respectively. When the inversion height is 2000 feet, the concentration will be, of course, half as much, that is, 1/2 part and 1/4 part per million, respectively. During the year we move up and down this line, or in the neighborhood, as indicated in the striped area of figure 3.9.

Several laboratories have investigated the degree of eye irritation at various concentrations of hydrocarbon and oxides of nitrogen. The eye irritation is often measured by having the subject look into the fumigation room while the eye is protected from the pollutants by a small sliding door. At the start of the measurement the sliding door is moved away and the eyes exposed. The time is noted between first exposure and onset of eye irritation. For severe irritation this may take from twenty to forty seconds. Low-level eye irritation takes as long as 120 to 140 seconds. Such experiments are not as easy as they appear as there is a large variation in the individual sensitivities of the subjects. Also, the individual may differ substantially on different days. Nevertheless, by carefully screening out the emotional and overcooperative, a group of reliable observers is obtained that can be used to establish the relation between eye irritation, hydrocarbons, and oxides of nitrogen levels. A series of tests are run at different concentrations of the pollutants and the relative intensity of irritation is determined. When the degree of eye irritation is known at enough points, it is possible to construct a three-dimensional graph relating eye irritation and the concentrations of oxides

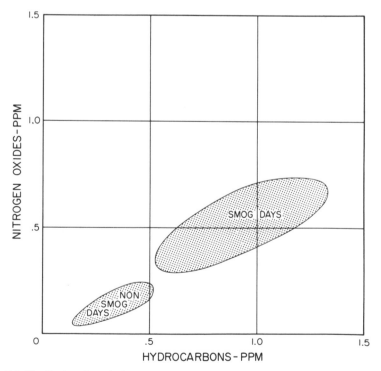

Fig. 3.8 Monitoring data during non-smog and heavy-smog days, Los Angeles County.

of nitrogen and hydrocarbons. Such graphs have been prepared by employees of the Taft Center of the Public Health Service (fig. 3.10).

Another way of presenting the same data is by drawing iso-irritation lines as in figure 3.11. These lines divide the irritation area roughly in regions of no, light, medium, and severe irritation. When one realizes the large spread in the responses of people, the artificial way of exposing the subjects, and the relatively large errors in establishing the various concentrations, it is clear that the nicely drawn lines in figure 3.11 are actually broad boundaries.

Now let us combine all three graphs: fumigation data, monitoring data, and source testing data. We see in figure 3.12 that there is a near perfect agreement.

The *no smog area* falls in the *no irritation area* of the fumigation experiment. The smog concentration coincides with the medium and severe smog and all data fall within the expected range of hydrocarbons:oxides of nitrogen ratio. When I decided to match these data, I did not expect very good agreement. This result exceeded my expectations.

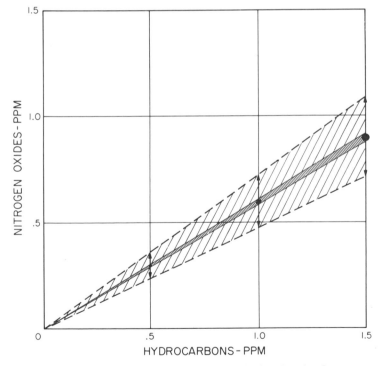

Fig. 3.9 Oxides of nitrogen/hydrocarbon ratio, Los Angeles County.

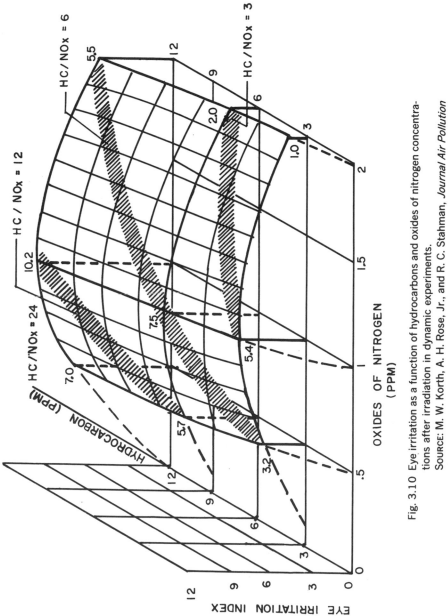

Fig. 3.10 Eye irritation as a function of hydrocarbons and oxides of nitrogen concentrations after irradiation in dynamic experiments.

Source: M. W. Korth, A. H. Rose, Jr., and R. C. Stahman, *Journal Air Pollution Control Association* 14 (1964): 168.

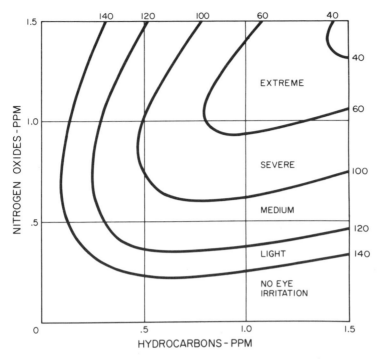

Fig. 3.11 Iso-eye irritation areas in the photochemical oxidation of auto exhaust.

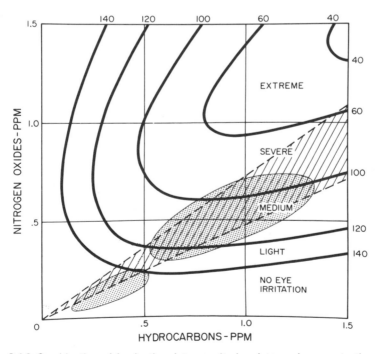

Fig. 3.12 Combination of fumigation data, monitoring data, and source testing data.

CONTROL POLICY

Credit should be given to the County Air Pollution Control District and the California Health Department for gathering the numerous data that are used in these graphs. The close correspondence of the three sets of data gives confidence that we can now go one step further and use these data as a basis for a program of control. To control the smog we have to bring the smog area down to the smog-free area, and according to the chart this means that we have to reduce both hydrocarbons and oxides of nitrogen by 70 percent.

There are some alternatives. One is a reduction of hydrocarbons alone; this would require a 95 percent reduction. The other possibility is a reduction of oxides of nitrogen alone by about 70 percent. The disadvantage of such proposals is that we end up with atmospheres rich in either hydrocarbons or oxides of nitrogen. Neither possibility appeals to me. One smells like a refinery, the other like a nitric acid factory.

Besides being an objectionable factor in the photochemical reactions, the nitrogen oxides are objectionable in their own right and are quite toxic even in low concentrations. On this basis, California's State Health Department has set community health standards which require a 75 percent reduction of the oxides of nitrogen in automobile exhaust as soon as practical devices are available.

Consideration of Los Angeles conditions is of far more general importance and applies in some degree to every modern metropolitan area where the automobile has become the major offender. The only difference is that the concentrations of the smog constituents (as in figure 3.8) would not be as high yet in other urban centers as in the confined, overpopulated Los Angeles area.

The present accelerated control efforts, especially those on automobiles, will be of decided benefit; with the explosive growth of our urban areas, however, it will be difficult to retrace our steps and achieve the clean skies of the past.

THE FUTURE

There are many who raise serious doubt about our ability to cope with the merciless increase in population and the consequent increase in pollution sources. One such individual is Dr. Philip Leighton of Stanford University. He writes:

In a sense air pollution may be likened to a weed. Controls may clip back the weed but will not keep it from growing again. To kill the weed we must get at the root, and the root of the whole problem of general pollution is combustion. . . .

The proper approach to a lasting solution of these problems, the only way to kill the weed, is to attack not the products of combustion, but combustion itself, to reduce by every possible means the burning of fuels in favor of non-polluting sources of heat and power . . . replacing fossil fuel by nuclear fuel, electric instead of hydrocarbon heat, in the home and industry, revamping our transportation systems and habits, and replacing the internal combustion engine by a non-polluting power source. Many of these changes are permanent assets to better living. None is beyond our technical competence. The burden falls on our social competence. All must be fought for; they will not come of themselves, and the fight will require both vision and courage. Whether or not we find the courage, the path is clear. We may be sure that only by such steps will we escape an unending procession of ever-increasing, ever more restrictive, ever unsatisfactory control.[5]

We have reached a turning point where the three major pollutants—hydrocarbons, oxides of nitrogen, and carbon monoxide—are decreasing, but are we going to reach the acceptable air quality levels proclaimed by the Federal Environmental Protection Agency? Calculations based on monitoring data in the Los Angeles Basin indicate that we cannot tolerate more than 200 to 400 tons per day of oxides of nitrogen and 1,000 to 1,500 tons of hydrocarbons.

The projected control of the hydrocarbons will almost meet this requirement by 1990 (fig. 3.13).[6] However, our technical ability to control the oxides of nitrogen falls far short of this goal. The projected control in 1990 will be about three to four times this figure (fig. 3.14).[6]

These figures show that *control technology alone* is not likely to cope with the ever-increasing growth in population and all of its polluting activities. We realize now that direct engineering control, necessary as it is, is only treating a symptom. It is like taking an aspirin for a chronic headache. The real trouble has deeper roots. It is an exciting and hopeful sign that we are now questioning for underlying causes of our predicament. Plans must be made *now* for how we want our world to look and the quality of air we want to breathe in the future. These plans of action must include conservation of green areas, development of a satisfactory transportation system, and management of our industrial complex with an overriding regard for its effect on the environment.

We are now critically examining the dogmas of the past: Is growth the indication of prosperity? Is growth the ideal that we should strive for? Are more people, more goods, or more wastes progress? Or do we put the emphasis on improving *the quality* of our life?

Such questions were taboo only a few years ago. Today we can raise them without being branded radicals. But will we be able to convert those who believe in the status quo, or those who blame everyone else other than themselves, or those who ignore the problem and hope it will disappear? Let me assure you it will not! We need all the persuasion and all the force

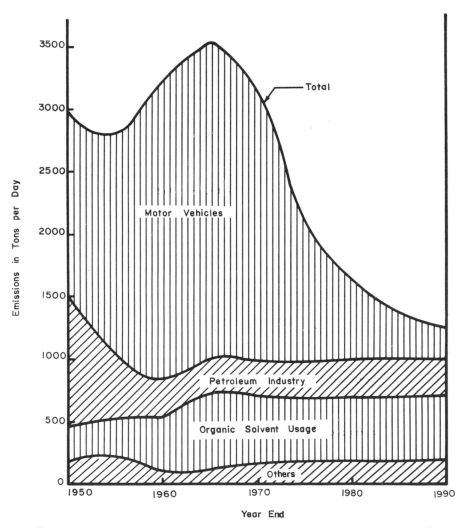

Fig 3.13 Estimates of emissions of hydrocarbons and other organic gases, South Coast
Basin.

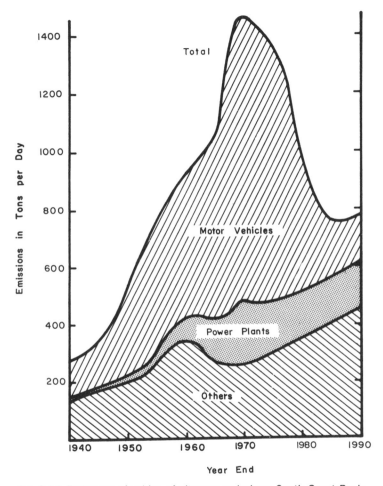

Fig. 3.14 Estimates of oxides of nitrogen emissions, South Coast Basin.

we can muster to plan for the future of our metropolitan areas. Every day we wait, new irreversible decisions are made that destroy more of our resources, our land, air, or water. There is no time to lose.

Reference notes

1. R. L. Duprey, *Compilation of Air Pollutant Emission Factors,* United States Department of Health, Education, and Welfare, Public Health Service, Publication no. 999-AP-42 (1968).
2. A. J. Haagen-Smit and L. Wayne, "Atmospheric Reactions and Scavenging," *Air Pollution* 1, ed., A. C. Stern (1968).
3. J. L. Mills, et al., *Emissions of Oxides of Nitrogen from Stationary Sources in Los Angeles County,* Los Angeles County Air Pollution Control District, Report no. 4.
4. State of California Air Resources Board, *Control of Vehicle Emissions After 1974* (Sacramento, Calif., November 1969).
5. P. A. Leighton, *Man and Air in California* (Statewide Conference, "Man in California 1980," Sacramento, Calif., 1964).
6. *Air Pollution Control in California 1970* (State of California Air Resources Board 1970 Annual Report to the Governor, January 1971).

4. Water Use and Management Aspects of Steam-Electric Generation

Victor A. Koelzer and Richard C. Tucker
National Water Commission, Engineering and Environmental Sciences Division

It will be our purpose to provide a broad perspective on the water use and management aspects of steam-electric generation. The projected requirements for cooling water are of growing concern. The Federal Power Commission has estimated that by 1990 withdrawals of water for condenser-cooling may be approximately one-sixth of the total average flow of all the rivers in the conterminous United States.[1] The Sport Fishing Institute reports that ". . . more ominously, during the two-thirds of the year when flood flows are generally lacking, about half the total freshwater runoff will be required for steam-electric-stations cooling water purposes at inland locations. . . ." [2] While these gross figures do not pinpoint specific problem areas, they do serve to indicate the magnitude of the problem on a national basis.

It also is well known that discharge of large quantities of heat in cooling water could have severe effects on the ecosystem into which the heat is dissipated. Increased temperatures and temperature changes can have lethal and sublethal effects on aquatic life. Reproduction, growth, and behavior may be seriously impaired or altered due to changes induced by elevated water temperature. Chemicals used to facilitate the cooling process and the design of related facilities could cause disruption in established, delicate ecological balances.

Before continuing, however, it will be useful to develop a framework within which these aspects may be properly analyzed.

Water resources must be viewed in the overall context of their interaction with the goals and objectives to which this or any nation aspires. In the past many have viewed the water resource as something to use rather than manage. Fortunately this is no longer generally accepted. Also water needs have frequently been extrapolated as a projection of past trends. Today we realize that future needs are dependent, to a great extent, on those policies which we choose today, both for quantity and quality.

Recent policy has clearly enunciated this nation's concern for a quality environment. Although greater focus has been given the great outdoors, the so-called natural environment, it is necessary to look at environment as encompassing those aspects comprising the entire human environment —sailing, air conditioning, nature trails, and, yes, the electric toothbrush. Perhaps in no other area of concern is the environmental issue more divisive than in the energy-environment issue, where developmental and environmental values directly confront each other.

ELECTRIC POWER GENERATION NEEDS

At the root of the energy-environment issue are the electrical power generating requirements of this nation. A number of forecasts of future electric power needs have been made by responsible bodies during the 1962–68 period. A comparison is given in table 4.1.[3] Although this tabulation indicates a certain amount of divergence in the later years, a large increase in electrical generation requirements was forecasted by each.

It is important to note that although a great deal of study is needed to better determine the energy use patterns in all sectors of the economy, the more broadly based forecasts are not mere percentage projections. Some discussion of the background of the recent forecast in support of the 1970 National Power Survey (NPS) by the Federal Power Commission (FPC) would be helpful to illustrate this point.

The NPS survey was based on assistance from six regional advisory committees representing all segments of the power industry. An appreciable amount of analytical detail was included in its derivation. Such breakdowns as farm and nonfarm, commercial, street-lighting, various industrial uses, and other uses of electrical energy were included in their analysis. In the residential sector, for example, growth in the use of nineteen classes of appliances was considered individually, by regions, and including consideration of saturation levels.

For our purposes, however, there is no need to judge the relative accuracies of any of the above forecasts. The principal point to be made is not that any one forecast is correct; it is that, regardless of the forecast, it appears the nation is faced with greatly increased demands for electrical energy. Most forecasts indicate a doubling by 1980—the lowest forecast shown in table 4.1 predicts that loads in the year 2000 will be three

TABLE 4.1
Utility Electric Power Generation
(billions of kilowatt-hours)

Forecast by	Date Report Published	Year for which Projection Was Made					
		1970	1975	1980	1985	1990	2000
NPS*	1964	1,484	2,024	2,693			
RAF	1963	1,287		2,084		3,044	4,467
NF&ES	1962			2,700			
CNP	1962, 1967			2,700			
CGAEM	1968	1,448	1,995	2,581	3,363		
PEC	1962			2,739			
EUS	1967			3,086			
EMUS	1968			2,739			
PCCP	1968			2,641			5,874

NOTE: Does not include industrial self-generation. NPS estimated this at 127 in 1980 for total generation of 2,820.

* Origin of individual forecasts is given at the end of this paper.

References Used in Table 4.1

CGAEM "Competition and Growth in American Energy Markets, 1947–1985," Texas Eastern Transmission Corporation, 1968.

CNP "Civilian Nuclear Power—A Report to the President," United States Atomic Energy Commission, 1962, and 1967 supplement.

EMUS "An Energy Model for the United States Featuring Energy Balances for the Years 1947 to 1965 and Projections and Forecasts to the Years 1980 and 2000," Bureau of Mines, IC 8384, United States Department of the Interior, July 1968.

EUS "Energy in the United States, 1960–1985," Michael C. Cook, Sartorius & Co., September 1967.

NF&ES "Report of the National Fuels and Energy Study Group on Assessment of Available Information on Energy in the United States," Committee on Interior and Insular Affairs, United States Senate, September 1962.

NPS "National Power Survey," Federal Power Commission, United States Government Printing Office, 1964.

PCCP "Projections of the Consumption of Commodities Producible on the Public Lands of the United States 1980–2000," prepared for the Public Land Law Review Commission, Robert R. Nathan Associates, Inc., Washington, D.C., May 1968.

PEC "Patterns of Energy Consumption in the U.S.," William A. Vogely, Division of Economic Analysis, Bureau of Mines, United States Department of the Interior, 1962.

RAF "Resources in America's Future," Landsberg, Fischman, and Fisher, Resources for the Future, Inc., Johns Hopkins Press, 1963.

and one-half times the 1970 load. *The general agreement is that large power load growth is inevitable.*

Sensitivity of Forecasts to Causal Factors

The forecasted growth in electrical energy consumption is linked to two seemingly inevitable trends: a higher population and increased per capita energy consumption. Despite the great concern about population growth, the most significant factor in changes in energy consumption is in per capita consumption. A recent statement by Resources for the Future points this out quite succinctly: "Ninety percent of the growth in power generation in the last thirty years has been caused by higher per capita consumption and only 10 percent by population growth." [4] This tells what has happened in the past. As for the future, the following quotation concerning the impact of per capita consumption in the Northwest is typical: "Economists compute that even if the objective of two children per family is achieved by 1975 (unlikely), it will not be until the year 2050 that our population will be stabilized." [5] Despite this, however, they go on to say: "Only 1.5 percent of the estimated 6 percent load growth per year in the next 20 years is attributable to population growth. The rest will come from increase in per capita consumption." There is no reason to assume that a similar situation will not be the case throughout the nation.

Many questions have been raised in recent years about the "need" for all the power that is being used. It has been suggested that the nation should restrict use of electric appliances, air conditioning, or other "luxuries" that utilize electrical energy. Others have taken different points of view.

It is not our function to suggest the policy the nation should choose in this respect, but it might be helpful to put the problem into perspective by indicating the forecasted uses for typical categories, as given in the 1970 National Power Survey for the South Central Region. This is done in table 4.2. It indicates in which use category the real reductions must be applied in order to have any significant effect on future electrical power requirements.

TABLE 4.2
Estimated Electrical Energy Requirements for South Central Region, 1990

Use Category	Projected Electrical Energy Requirement–1990 (million kwh.)
Farm (excluding irrigation and drainage pumping)	8,613
Irrigation and Drainage Pumping	984
Nonfarm residential	242,760
Commercial	134,235
Industrial	418,963
Street and highway lighting	4,265
All other	18,355
Losses and energy unaccounted for	72,205
Total	900,380

SOURCE: Federal Power Commission, "Problems in Disposal of Waste Heat from Steam-Electric Plants" (Washington, D.C., 1969), p. 38.

It is enlightening to consider the effect of restricting or eliminating various uses in the home to further develop this perspective. The detailed breakdown for the nonfarm residential use category as given by the FPC in support of table 4.2 supplies the needed information. If, for instance, we were to assume that freezers, dishwashers, food waste disposers, and clothes dryers were nonessential and should be banned, the total electrical energy projection for the South Central Region in 1990 would be reduced only from 900,380 million-kwh. to 885,170 million-kwh., or less than 2 percent. If, in addition, all home air conditioning was also considered as a luxury and could be banned, the projections would drop to 834,420 million-kwh., giving a total reduction of approximately 7 percent. These figures indicate that reductions in these items of home use will not be highly effective in reducing overall load growth.

It is clear that mere restrictions on the use of a few appliances will not solve the energy-environmental issue. Notions of reduced electrical energy requirements must entail a deeper analysis of this nation's aspirations for economic growth, the role of energy in ensuring the quality of the total environment, future pricing policies, and the desired future life-styles of the population.

Sources of Electrical Energy

A recent statement by the chairman of the Federal Power Commission included estimates of future sources of electrical energy.[7] These figures are given in table 4.3.

The sources and amounts of waste heat discharged to the nation's waterways in the future will be directly related to the mix of types of steam-electric generating plants in use. The decisions affecting the type of plant and fuel in use will be governed by a number of factors, only one of which will be thermal discharges.

TABLE 4.3
Projection of Generating Capacity

	1970		1980		1990	
	Mega-watts	Percent-age	Mega-watts	Percent-age	Mega-watts	Percent-age
Conventional hydro	51,700	15.2	68,000	10.4	82,000	6.5
Pumped storage hydro	3,600	1.1	27,000	4.0	71,000	5.6
Fossil steam	260,300	76.5	393,000	59.0	557,000	44.6
Internal combustion and gas turbine	18,300	5.4	30,000	4.5	50,000	3.9
Nuclear	6,100	1.8	147,000	22.1	500,000	39.4
Total	340,000	100.0	665,000	100.0	1,260,000	100.0

The figures in table 4.3 indicate that industry presently expects that the greatest impact will continue to be from steam power plant operation, but with a trend away from fossil-fuel generation toward the less thermally efficient nuclear capacity. Similar views are given in the following excerpts from a recent report by the Office of Science and Technology (OST), which was based on forecasts similar to those given by Chairman Nassikas.[8]

TABLE 4.4
Estimated Number of Thermal Generating Plant Sites
(500-megawatt capacity and above for year 1990)

NPS Region	Fossil-Fuel Plants	Nuclear Plants
Northeast:		
Total sites	41	45
New sites	5	38
Cooling towers	8	9
Southeast:		
Total sites	34	60
New sites	6	45
Cooling towers	4	32
East Central:		
Total sites	62	21
New sites	16	17
Cooling towers	17	6
South Central:		
Total sites	92	22
New sites	49	22
Cooling towers	27	5
West Central:		
Total sites	25	19
New sites	6	11
Cooling towers	6	8
West:		
Total sites	38	33
New sites	9	31
Cooling towers	21	15
Total United States:		
Total sites	292	200
New sites	91	164
Cooling towers	83	75

The OST report states as a footnote to this table:
Estimates are based on preliminary information assembled by FPC staff in connection with work in updating the National Power Survey. The staff of the Water Pollution Control Administration in the Department of the Interior, in reviewing the data, suggests that the number of plants requiring cooling towers may be greater than the FPC staff estimate.

. . . 255 new sites would be needed (by 1990) for thermal plants of 500-mw capacity and larger. Of this total, 91 sites would be for fossil-fuel plants and 164 would be for nuclear plants.

. . . the cooling requirements for adequate waste heat disposal will severely limit the acceptable locations. Although the majority of the new plants are in the 1,000- to 4,000-mw range, a significant number exceed 3,000-mw. The largest closely approaches 10,000-mw.

Table 4.4 gives, by regions, the OST estimates of the number of fossil-fuel and nuclear plants through 1990.[8] It is in the siting of these plants that possible environmental degradation becomes an issue. In fact, the footnote to the table clearly indicates a difference of opinion between two federal agencies as to the need for cooling towers.

PERSPECTIVE ON WATER AVAILABILITY AND NEEDS

National Picture

First let us look at the nation's water resources in relation to all uses of water. A number of forecasts of water needs and supplies have been made in the past, and these forecasts have varied widely. For example, in 1961 the Senate Select Committee showed apparent deficiencies of supply versus demand in eight of twenty-two regions of the United States by the year 2000. Four years later, the United States Geological Survey, using the same basic data, but applying different basic assumptions, showed apparent deficiencies in nineteen of the regions. Both were correct in their data, but used different assumptions to arrive at their figures.

This demonstrates the hazard in attempting to simplify, for popular use, an inherently complex relationship. Regional aggregations on areas as large as the twenty-two regions are frequently misleading. In areas of water scarcity, comparisons of available supply with water needs become meaningful only when we localize both factors, and then only if we look at the supply in terms of the amount of water that is used upstream and returned to the river. In other words, it is the return flow, or reuse capability, that makes most regional aggregations misleading.

We know, of course, that there are pockets of water scarcity in the nation. Arizona and the Pacific Southwest represents one of the major areas. It does not necessarily follow, however, that the region must curtail industrial growth because of lack of water. New sources of water, such as seawater (desalted if necessary), precipitation modification, geothermal sources, and interbasin transfers represent possibilities. There are many economists who say that it would be much better to transfer water presently used for irrigation to higher valued municipal and industrial uses, rather than turn to the new sources that have been mentioned. Some engineers and economists also suggest there are other steps we could take that would increase efficiency in the use of present supplies. All of these alternatives deserve investigation.

General Role of Cooling Water Needs

Forecasts of the amount of cooling water the nation needs are, quite obviously, highly dependent on the type of cooling used at the generating station. This will be discussed later. In addition, however, forecasts of cooling water needs can be highly misleading if the difference between withdrawal requirements and the amount of water actually consumed are not made clear.

We can state quite accurately, for example, that by 1980 water withdrawals for cooling purposes, nationwide, can be expected to surpass those for agriculture. The Water Resources Council's estimates of withdrawals by 1980 are 215 million acre feet (MAF) for cooling (66 MAF from saline sources), as compared with 152 MAF for agriculture. However, it should be noted that the Council's estimate of consumptive use for cooling in 1980 is only about 2.5 MAF, compared with consumptive use of 91 MAF for agricultural uses.

QUANTITY AND QUALITY ASPECTS OF COOLING WATER USE

Heat Rejected by Typical Plants

The type of generating facility, plant heat rate, and the type of cooling method used are the overall factors involved in determining the quantity of cooling water needed and the quantity of heat that will be released to water bodies.

Table 4.5 presents a comparison of the heat and cooling water characteristics of typical steam-electric plants of different design. While discussion of a "typical" plant is risky, that risk is considered necessary to obtain a perspective of the potential impact of different types of plants.

Thermal Efficiency

The thermal efficiencies of modern fossil-fuel and nuclear power plants generally range between 33 and 40 percent. In other words, 60 to 67 percent of the heat generated is nonproductive in the generation of electricity and is therefore rejected to the biosphere as waste heat. A small portion of this

TABLE 4.5
Heat Characteristics of Typical Steam-Electric Plants
(heat values in Btu. per kwh.)

Plant Type	Thermal Efficiency (percentage)	Required Input (heat rate)	Required Input Minus Heat Equivalent *	Lost to Boiler Stack † (etc.)	Heat Discharged to the Condenser	Cooling Water Requirement (cfs. mw. of capacity) ‡
Fossil-fuel	33	10,500	7,100	1,600	5,500	1.6
Fossil-fuel	40	8,600	5,200	1,300	3,900	1.15
Light water reactor	33	10,500	7,100	500	6,600	1.9
Breeder reactor	42	8,200	4,800	300	4,500	1.35

SOURCE: Federal Power Commission, "Problems in Disposal of Waste Heat from Steam-Electric Plants, June, 1969" (Supporting the Commission's *National Power Survey*).

* The heat equivalent of one kilowatt-hour of electricity is 3,413 British thermal units (Btu.).
† Approximately 10 to 15% x required input for fossil fuel.
   Approximately 3 to 5% x required input for nuclear.
‡ Based on a temperature rise of 15° F.

is discharged to the air, as indicated in table 4.5, but the majority is released to water bodies.

In contrast to the very large growth in electrical power production, there have been only modest increases in thermal generating efficiency over the past fifteen years. Little that we do today is expected to have any appreciable impact on thermal efficiencies of plants constructed through the next several decades. Although breeder reactors will increase efficiency of nuclear plants, probably by the late 1980s, the increase will still not be significantly beyond the 40 percent efficiency that represents one end of the current range. It is especially interesting to note from table 4.5 that the quantity of heat released to the condenser for the breeder reactor, and therefore the cooling water requirement, is greater than for modern fossil-fuel plants. In other words, breeder technology will not significantly change the heat that will be discharged from the generation process.

It is conceivable that twenty to thirty years of concerted effort could bring in new forms of generation, such as magnetohydrodynamics, the fuel cell, geothermal energy, or nuclear fusion, that would effect significant reductions in heat discharges. It will, of course, take radically different degrees of effort to bring these into being. There is little agreement, moreover, as to when and how this might take place.

It is thus apparent that there is a need for two perspectives. One perspective must deal with the near to intermediate term, a period of time during which power plants must be sited based on currently generally known and available technology. This time period is expected to include the remaining part of the twentieth century. The second perspective must deal with that time beyond when research and development efforts might have a significant effect on the whole issue of energy generation and waste heat discharge.

Effects of Various Cooling Methods

The amounts of water consumed and heat carried from the condenser to the water body are totally dependent on the type of cooling mechanism used. Conventional cooling systems, which depend primarily on evaporative cooling to dissipate heat, cause consumptive losses which, in many areas, could be quite significant. These losses are on the order of 1 percent of condenser flows for once-through cooling, 1.5 percent for cooling ponds, and approximately 2 percent for evaporative cooling towers. If dry cooling towers are employed in the future there would be no consumptive loss and no discharge of water, since for cooling these towers depend on the convective transfer of heat to the air as it passes through a fin-tube heat exchanger.

There is also a great deal of concern about the possible adverse ecological, climatological, and esthetic impacts of these various cooling techniques. Some concern is expressed through purely subjective evaluation, such as the esthetic effects of cooling towers on the landscape. Other aspects are subject to quantitative analysis, particularly with respect to cost. Perhaps the greatest divergence of opinion occurs on the effects of cooling towers on local climate through induced fog, precipitation, and icing.

Obviously the type of cooling mechanism employed will be a function of the water quality standards for the receiving water body and the economics of the cooling systems available for use.

Cooling Water Needs

As part of the background studies for the National Power Survey, the FPC has prepared estimates of potential waste heat output and related cooling needs through 1990.[9] Four methods of cooling for removal of waste heat from electric power generation were considered. Three of these involved use of freshwater: once-through cooling, cooling ponds with freshwater, and cooling towers of the evaporative type. The fourth method was once-through cooling using saline water. Three planning assumptions were tested:

Assumption A. Each plant would be expected to utilize the cooling water system having the lowest cost and causing the least physical intrusion on the environment, provided that the resulting operation would not violate approved local water temperature criteria.

Assumption B. Each plant or plant addition constructed after 1970 would be expected to require auxiliary cooling by a cooling pond or an evaporative cooling tower unless it could utilize water from the ocean, with a long outfall.

Assumption C. Each plant operating in 1980 and 1990, regardless of when constructed, would be expected to require a cooling pond or an evaporative cooling tower unless it could utilize water from the ocean, with a long outfall.

The study was made for each of the six regional advisory committee areas. The results for the entire United States are given in table 4.6. These results, in effect, show the influence of different assumptions concerning acceptable environmental standards, with particular emphasis on the quantities of water consumed.

It should be noted that under any of the assumptions in the FPC study the quantities are significantly greater than estimated by the Water Resources Council in 1968. This undoubtedly reflects, in part, assumptions of greater use of evaporative cooling towers, which have higher evaporative losses, in the more recent study reflected in table 4.6. This is one measure of the impact of mounting environmental concern on our assumptions regarding future water needs.

TABLE 4.6
Estimates by Federal Power Commission Staff Cooling Facility Requirements

Category	Year	Assumption A	Assumption B	Assumption C
Plants using	1970	23	—	—
cooling ponds	1980	59	100	128
(number of plants)	1990	76	135	154
Plants using	1970	43	—	—
cooling towers	1980	123	176	264
(number of plants)	1990	181	273	348
Water consumption	1970	1.02	1.02	1.02
(million acre	1980	3.08	4.18	4.75
feet/year)	1990	7.27	10.02	10.64
Investment in	1970	814	814	814
cooling facilities	1980	2673	3121	4831
($ millions)	1990	6265	7983	9767

NOTE: Figures represent cumulative totals to the year indicated.

Effects of Thermal Discharges on Water

The discharge of heat to a body of water can cause a number of physical and biological changes. As water temperature is raised, the oxygen holding capacity of the water decreases, the reaeration rate rises, the density changes cause possible stratification, the chemical reactions tend to proceed at a faster rate, and the viscosity decreases which reduces the sediment-transporting ability of the water. These changes, of course, can be beneficial or detrimental, depending to a great extent on the way in which water in the receiving body is expected to be used.

There could be a variety of impacts. The treatment of public water supplies, for example, could require a smaller amount of chemicals if significant changes occur in the temperature of the receiving waters, since chemical reactions will occur at a faster rate. At the same time, however, the stream's waste-assimilation capacity could be lowered. The addition of heat in the winter months could lengthen the shipping season by shortening the period of ice cover in the shipping lanes. Discharge of heat in colder waters, such as along the Pacific coast, could lengthen the recreation season.

At the present time a great deal of emphasis is being placed on the potentials of a higher degree of planned beneficial use of heat discharges. These range from mariculture, aquaculture, and agriculture to district heating and air conditioning.* While these do have promise, to date they do not appear to offer much hope for reducing near-term problems at least on a broad scale. For the long term, however, continued development of beneficial-use technology, coupled with the creation of an institutional climate responsive to the application of such technology, might provide attractive alternatives.

The greatest potential impact of heat discharge to water is on aquatic life. A large number of studies have been and are being conducted on this subject. Although a great deal of additional in situ study must be made of specific species and their tolerance ranges, a number of basic relationships are fairly well understood. Temperature changes have a direct effect on such things as metabolism, reproductive cycles, behavior, and digestion and respiration rates. Adverse effects can be lethal as well as sublethal. Mortality can occur due to exceeding maximum temperature tolerances, as well as from rapid increases in temperature. A shift in population structure that affects specie diversity can have a substantial impact on ecosystem stability. This can occur throughout the food chain, from plankton, periphyton, filamentous algae, rooted aquatic plants, and invertebrates, to fish. Since fish are at the apex of the aquatic food pyramid, any drastic change in any

---

* Mariculture is used to differentiate production processes in saltwater, whereas aquaculture refers to similar processes in freshwater.

part of the pyramid will be rapidly reflected in altered fish population. The more we learn about these effects the more knowledgeable our decisions will be concerning plant siting and the environment.

Water Quality Standards

The relationship of the heat discharge problem to water quality standards is an involved topic. Although it is not within the scope of this chapter to give a complete coverage of standards, it is significant to point out that the setting of water quality standards is one of the key issues surrounding the whole issue of heat and its effects on the aquatic ecosystem.

Embodied in the technical input to standard setting are such things as maximum permissible temperatures, maximum changes in temperature, and maximum rates of change of temperature, all of which should vary according to the body of water as well as geographically and from season to season. It is worthy of note that for streams expected to have more than one use, the criteria applicable to the most sensitive use would take precedence in establishing standards. In most cases, the criteria applicable to fish and other aquatic life would be controlling.

Other Water Related Aspects of Power Plant Operation

Various other aspects not directly concerned with heat but intimately related to the operation of condenser cooling systems merit attention as they tend to be forgotten in the broader studies on power plant siting.

Chlorine is sometimes added to the cooling water supply to prevent fouling of the condensers. This can have an adverse effect on fish as well as those organisms in the food chain which are circulated in the cooling water system.

A number of chemicals such as sulfuric acid, calcium sulfate, chlorine, polyphosphates, chromates, and other corrosion inhibitors have been used to treat the water circulated in a cooling tower.[10] The effects of these chemicals, especially with respect to possible synergistic effects with heat additions, must be given careful consideration.

Metals have also been identified as affecting aquatic life. In one case water chemistry data showed increased copper concentration in the effluent water from a steam-electric plant on the Patuxent estuary.[11] Even after the corroded condenser tubing was replaced by a more resistant metal alloy, green oysters, presumably caused by residual copper concentrations, were harvested in the surrounding areas.

Mechanical and hydraulic effects can also be detrimental to the environment and must be given careful consideration in the site selection and design of generating and cooling facilities.

OVERVIEW OF IMPORTANT FACTORS

These comments provide a broad overview of those aspects of water resources which are of most concern with respect to steam-electric power generation. It is relatively easy to identify those factors of greatest importance: population growth, per capita consumption of electrical energy, means of generating electrical energy, means of cooling or using waste heat if any is produced, and the quantitative and qualitative effects of this heat on the water resource. Although we are faced with severe technical problems, the most difficult task is to identify the type of total environment which we as a nation choose for our future—to achieve a balance between so-called environmental values and developmental values. As one water official recently put it, "Although everyone agrees there should be a balance, the fighting is precisely over what 'balance' is to mean—where balance is to be established, what's to go into it, in what proportion and priority."

Reference notes

1. Federal Power Commission, "Problems in Disposal of Waste Heat from Steam-Electric Plants" (Washington, D.C., 1969), p. 38.
2. *Sport Fishing Institute Bulletin,* no. 191 (Washington, D.C., January–February 1968): 1.
3. Pacific Northwest Laboratories of Battelle Memorial Institute, "A Review and Comparison of Selected United States Energy Forecasts" (Study prepared for the Energy Policy Staff of the Office of Science and Technology, Executive Office of the President, December 1969), p. 26.
4. *Resources Newsletter,* no. 34 (Washington, D.C.: Resources for the Future, June 1970): 1.
5. *Bulletin of the Northwest Public Power Association,* August 1970.
6. Federal Power Commission, "The 1970 National Power Survey," pt. 3, sec. III-1.
7. J. N. Nassikas, "An Analysis of the Current Energy Problems" (Speech before the Electrical World Conference for Utility Executives, Washington, D.C., 14 January 1971), p. 11.
8. Energy Policy Staff, Office of Science and Technology, "Considerations Affecting Steam Power Plant Site Selection" (Washington, D.C., December 1968), pp. 4–5.
9. Frederick H. Warren, "Electrical Power and Thermal Output in the Next Two Decades," in *Electric Power and Thermal Discharges,* Merril Eisenbud and George Gleason, eds. (New York: Gordon and Breach Science Publishers, Inc., 1970), pp. 21–46.
10. I. D. Kolflat, "Hearings before the Subcommittee on Air and Water Pollution of the Committee on Public Works," United States Senate, Ninetieth Congress (Washington, D.C., 6 February 1968), p. 64.
11. J. A. Mihursky, "Patuxent Thermal Studies Summary and Recommendation," Natural Resources Institute, University of Maryland, NRI Special Report no. 1 (1969), p. 5.

5. Central Station Energy Requirements

F. A. McCrackin
Southern California Edison Company

The surging concern about the degradation of our environment has focused public attention on estimates of future energy requirements. In certain quarters, questions have even been raised on whether the expansion of our consumption of electricity is not getting out of hand. Simultaneously, fears have been expressed by others that power shortages are imminent in some areas because of low reserve margins and generating capacity.[1]

There is a "surging concern about the degradation of our environment" and there is a "surging concern" of public attention focused upon all forms of energy growth, especially on electrical energy growth. During most of this century, except for a pause in the depression years, a rapid rise in the demand for electricity has occurred. United States electrical demand has doubled roughly about every ten years in response to an increasing population and an increasing consumption of energy by that population. Will this trend continue, and can we afford such a trend? These are questions often asked of utilities by a concerned public.

I do not think that United States energy requirements will continue indefinitely to double every ten years; in fact, I expect energy requirements will continue to increase for some time but at a declining rate. That is, the rate of increase will decline. Indeed, we cannot tolerate a continuous course of unlimited energy growth in view of present and projected environmental ills.

To gain a further understanding of the energy growth issue, let us briefly review the methodology of load forecasting.

ENERGY GROWTH FORECASTING
METHODOLOGY

All forecasting methods appear to include the same fundamental steps although the rigor of their application is variable: (1) collecting historical data and adjusting or normalizing them to achieve reasonable consistency over the desired time period; (2) analyzing data to determine the significant economic, demographic, and geographic factors which have influenced load growth in the past; (3) extrapolating significant causal factors into the future and determining the degree of uncertainty in each extrapolation; (4) deriving forecasts by considering the effects that these extrapolated factors are expected to have during the time period chosen; (5) determining the relative degree of uncertainty in the forecast; and (6) periodically comparing recorded performance with previous forecasts and making appropriate adjustments in techniques and assumptions.

Despite the many improvements being made in energy growth forecasting data and methods, there is little that is mechanical in effective forecasting. It is not useful to simply extrapolate from a historical growth curve, disregarding changing population trends, social values, and economic climates. Even the most sophisticated techniques do not eliminate the need for good judgment. However, such techniques can provide a better understanding of the past and, thus, a better base on which judgments can be made concerning the probable shape of the future.

Estimates of future demand and energy requirements constitute the foundation for planning in the energy industry. Every producer and distributor of energy, large or small, wholesaler or retailer, requires demand and energy forecasts upon which to base physical and financial planning. Others associated with the energy industry, such as suppliers, regulatory agencies, and appliance manufacturers, also need such forecasts to carry out their functions.

Capacity must be provided in the physical system to meet expected peak demand requirements. Demand forecasts for the product are the basis for planning additions to production, transmission, and distribution plants. In addition to maximum demand forecasts, the load shape (hour-by-hour demand estimates for the period) may influence the kind of production capacity, especially with the electric utilities who cannot store their product.

The combination of demand and energy forecasts forms a basis for planning fuel requirements.

Energy and demand forecasts are also the starting point for financial planning. The capacity and fuel strategies based on these forecasts are translated into financial requirements. Energy forecasts are the basis of revenue planning. Together these factors form a financial plan. In addition, they are used to make rate analyses and to develop comprehensive marketing plans.

FACTORS INFLUENCING ENERGY GROWTH

As noted before, the major factors influencing future needs for energy are population, land use, national and regional economics, technology, and the availability of natural resources. Social values, which have not changed dramatically in the past, have not yet played a major role in establishing energy demands. In some quarters, arguments are being raised that social values should be altered for the direct purpose of influencing energy demands.

Population

The number of people in a given geographic area has a direct impact on the energy needs for that area. This relationship also can be influenced by other factors, such as the efficiency of energy-using machines and appliances and energy use habits of a population.

The population growth of a region is affected by various restraints, such as availability of land, fluctuations in personal income, employment opportunities, and changing social values. Oddly enough, a study of fruit fly populations provides an analogy to assist the understanding of the effects of population constraints.

Figure 5.1 indicates that the rate of growth for an initial, small fruit fly population is slow, but as offspring are born and mature to become parents, the curve begins to climb abruptly. The resulting large population soon reaches a limit set by a resource, such as the availability of food per unit of time, and levels off (note dotted line area, fig. 5.1). If additional amounts of that resource are made available, the exponential growth begins again, climbing to Point B, where once again the resource is fully utilized. If the resource is again made available, the curve climbs to Point C, set by the new level of availability of that resource or by the limited availability of some other resource, such as space, air supply, etc.

Man faces similar limitations. He has a limited supply of resources in a given geographic area. When he has utilized an available but limited resource within an area, and when importation of that resource is not possible, population growth will level off. A good example might be water resources. When a population reaches a level which fully utilizes its water resources and no additional resources are available via importation, the ability of that population to expand further is constrained and growth must cease until additional supplies of water are available.

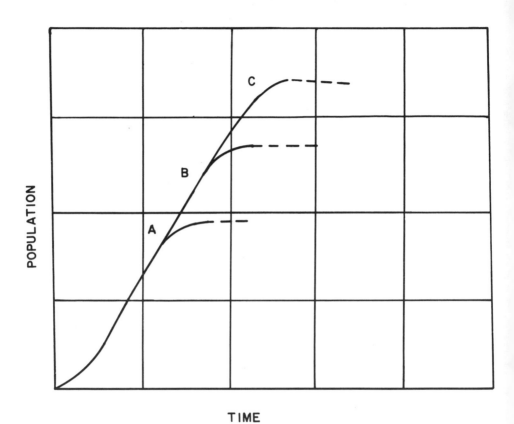

TIME

Fig. 5.1 Growth of small fruit fly population as a function of food availability.

The environmental implications of this analogy are obvious. If man does not eventually control population growth and develop the technology to conserve natural resources, population will be limited by natural causes. Unfortunately in the real world the natural sources of population control may prove to be catastrophic in effect: Population control through famine, pestilence, plague, and war is not pleasant to contemplate.

Land Use

Land use is a key factor affecting energy use. The present and zoned uses of land in a specific area may limit population expansion and the density of energy use. For example, an area may have a zoning designation that allows only single-family dwellings to be built. If city officials decide to rezone the area to allow the construction of multi-family dwellings, it is probable that energy use for that specific area will rise dramatically. Zoning, then, affects the population density of land and the density and character of energy use.

Economics

The general condition of our economy, both short and long term, has a direct influence on energy demands. This is evidenced by the dip in energy growth curves caused by the depression and post-depression "recovery" years, 1930–40. GNP growth generally indicates that industrial facilities are expanding their operations and increasing energy consumption. Additionally, expanding GNP results in an expanding employment market and has been generally accompanied by expanding personal income. Although gains in GNP are generally accompanied by increases in energy consumption, the energy increase is not necessarily in direct proportion to that of GNP. Factors tending to minimize the increment of energy required to achieve a gain in GNP are:

1. Increased thermal efficiency of prime movers—i.e., steam and gas turbines, internal combustion engines, etc.
2. Structural changes in the "product mix" of the economy
3. The refinement of industrial production methods
4. The rise of electricity as a source of energy

Technology

Historically, technological advances have been accompanied by an over-all reduction in energy use per unit of work or benefit, produced by increases in the thermal efficiency of energy using machines and devices. For example, the central furnace in a modern home is more efficient, in the sense of making available more of the fuel heat content, than a coal or wood stove, not to mention an open fire. Replacement of steam by diesel power in railroad locomotives resulted in a major improvement in the efficiency of fuel utilization. There has also been a constant increase in the efficiency with which fuels are converted into electricity. The latter example is not only the best documented, but perhaps the most spectacular on record: In 1900, almost seven pounds of coal were required to produce a kilowatt-hour of electricity; by the late 1960s, an average of approximately nine-tenths of a pound was required.[2]

Natural Resources

A relatively obscure stimulant to the increased use of energy is the diminishing quality of natural resources. In past centuries, man has skimmed the cream from the earth's wealth of natural resources and left us with resources of lesser quality. Larger amounts of energy will be required to convert existing resources into usable products.

Dwindling supplies of fossil fuels, such as coal, oil, and natural gas, create a stimulus toward increased utilization of nuclear fuels and of electricity as a means of transporting and using energy derived from nuclear sources. By the year 2000, supplies of gas are expected to be limited and supplies of oil are expected to be marginal; in addition, although coal has the potential to satisfy much of the energy demand in the year 2000, technological and economic problems associated with air pollution may restrict utilization of coal.

Social Values

Social values have not played a prominent role in stimulating energy demands in the past; however, current concern for environmental and urban ills may significantly alter social values and markedly influence energy demands. Two key factors should be considered.

First, within the United States there is the beginning of a shift from a growth ethic to a conservation and quality oriented society. This factor may tend to cool the economy and slow the energy growth curve.

Secondly, associated with this change in value orientation, public pressure is being exerted upon industry to minimize and eliminate all forms of pollution. In contrast, this factor will tend to increase energy consumption. More energy will be needed to produce and operate pollution control equipment. Additionally, efforts to cure urban ills through renewal programs will have a similar effect.

TOTAL ENERGY REQUIREMENTS

Total requirements for all energy forms in the United States are viewed here from two perspectives, national and regional.

National Growth Curve

Figure 5.2 shows that total energy growth is expected to increase by about one-fifth in each decade between now and the year 2000.[1] Figure 5.3 indicates that this growth will be influenced by gross national product.[3] Although imperfect, GNP is a fair measure of living standards. Economically developed countries, such as the United States and Canada, have higher levels of both energy use and living standards. The underdeveloped countries with poor living standards use practically no energy; they are depicted in the lower left-hand corner of the graph.

Population will also affect the energy curve. Like motherhood, America's growing population has long been deemed a blessing. Large families meant more hands on the farm, bigger markets for business, and security for aging parents. More people spelled progress and overpopulation was something that happened elsewhere. Now we realize that more people spell more population and overpopulation has become a matter of concern in the United States. Changing social values will tend to decrease the rate of growth for our population. A group known as Zero Population Growth (ZPG), founded by Stanford biologist Paul Erlich, is adding 1,000 new members per week; its slogan is "Stop at Two."

In theory, two children per family would exactly replace their parents. Since some will die before replacement, the total fertility rate needed to achieve replacement and stability is estimated to be 2.11 children per woman. A century ago the total rate was about six, and as recently as 1957 it was 3.77. By 1970 it had dropped to 2.45, not much above the replacement level. With the growing acceptance of contraception and abortion, there is a good probability that we, the United States, could decrease to the replacement fertility rate in this decade, according to Herman Miller, Chief of the Population Division of the Census Bureau.[4] Even if the replacement fertility rate is reached in this decade, total population will continue to grow at a slowing rate until a stable distribution in all age groups is attained. Population stability would not be reached for many decades, perhaps six or seven.

Federal legislation to limit income tax dependency deductions to no more than three dependents in a family, currently being proposed, would give further support to population growth stabilization.

As demonstrated in figure 5.4, an era of high birth rate generally has an impact approximately twenty years later in the form of another era of high birth rates when children from the previous era become parents. However, with expected changes in social values, the second lump on the curve will probably be smaller, because the children of the previous era might be conditioned to think: "Stop at Two."

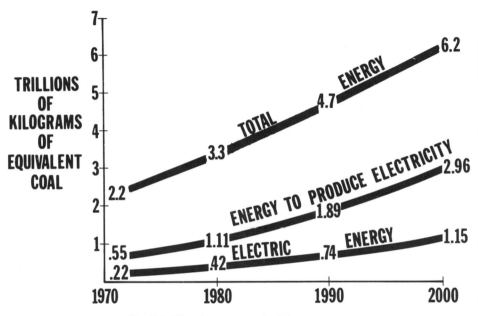

Fig. 5.2 Electric energy and total energy use.

104

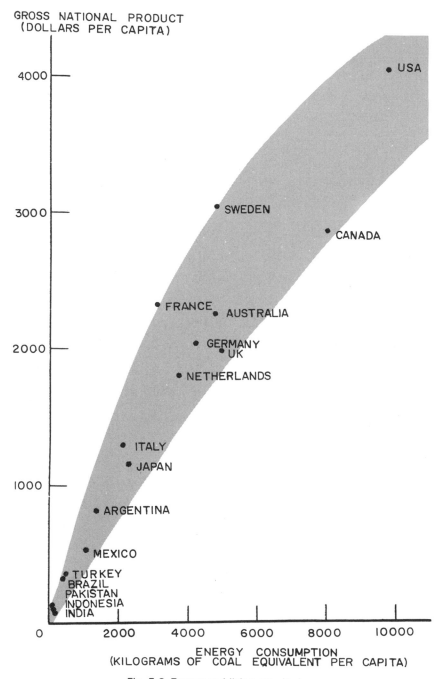

GROSS NATIONAL PRODUCT
(DOLLARS PER CAPITA)

Fig. 5.3 Energy and living standards.

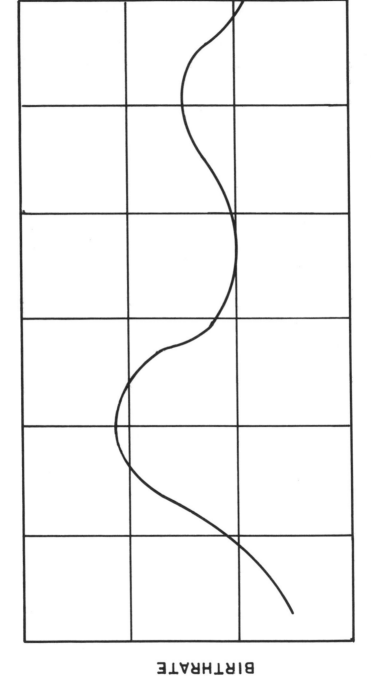

BIRTHRATE

TIME

Fig. 5.4 Idealized cyclic behavior of birth rate versus time.

In view of the current trend towards zero population growth, it is anticipated that energy growth rates will decline.

Technological advances, as thermal efficiencies are improved, are also expected to stimulate declining rates in energy growth. In contrast, the declining quality of our natural resources will tend to work against the effects of improved technology, requiring added increments of energy to produce the same quantities of goods.

Regional Growth Curves

Arizona, California, and several other western states, including the Pacific Northwest, are all part of the Federal Power Commission's West Power Region. Figure 5.5 indicates the 1970 National Power Survey forecast of total energy growth for this region. This forecast is based on several assumptions listed below:

1. United States population: growing at about 1.6 percent per year, reaching about 285 million by 1990
2. Gross National Product and constant dollars: up about 3.5 to 4 percent per year
3. Defense: at cold war level
4. Business cycles: fluctuations above and below long-run trend, with recessions similar to those after World War II, but no severe depression
5. Air and thermal pollution controls: increasing at an accelerating rate
6. Technology: gradual rather than revolutionary changes
7. Public policy: expected to change in response to environmental protection demands
8. General price levels: up about 1.52 percent per year, with fuel prices up slower and with long-run price level changes not being a decisive factor in the choice of fuels
9. Deviations from these assumptions: will tend to balance out [5]

As previously indicated, energy use is a measure of economic progress. In the rapidly growing West, the use of energy increased at an average of 4.4 percent per year during the period 1950 to 1965, compared with about 3.5 percent for the nation. The difference chiefly resulted from the more rapid population increase in the West, 3 percent per year as compared to 1.6 percent for the nation. In the future, western population growth is expected to slow. Since there is a relationship between population growth and energy use, energy use in the western area is also expected to grow at a slower pace.[1]

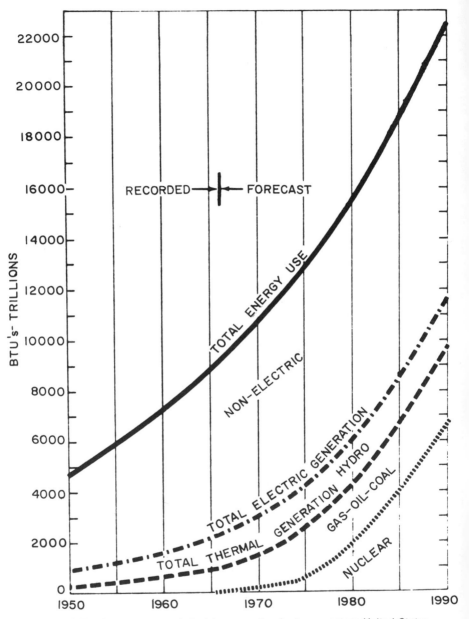

Fig. 5.5 Total energy use and electric generation by type, western United States.

ELECTRICAL ENERGY REQUIREMENTS

National Electrical Energy Growth

Many fear that electrical consumption will indefinitely continue to double every decade, as it has historically (see fig. 5.2). The simple fact is that electrical consumption cannot continue to double every ten years. One national forecast suggests that the annual rate of growth will be about 7 percent in 1970 and will decline to a value of about one-sixth less every ten years thereafter. Thus, the growth rate would drop to 6 percent in 1980 and 5 percent in 1990, etc.[7] A declining population growth rate, an eventual market saturation for energy-using devices in the home, and new social values tending towards conservation of energy all seem to indicate that the rate of growth for electrical energy will decrease, not increase, in the coming decades. Absolute increases in electrical energy use, however, will be very large.

The stimuli which will keep the electrical energy curve on a course of steady increase at a declining rate are already within our present horizon. Some of these stimuli are: (1) In transportation within cities, the greater use of electric-powered mass transit and the potential of battery-operated vehicles; (2) Increased use of high-speed electric trains for transportation in suburban and interurban services; and, (3) In chemical, petroleum, mining, and metal industries, electric energy will make possible new and improved extraction, refining, conversion, and pollution control processes. A significant use of electrical power will be in the recycling of waste produced in other sectors of the economy and in eliminating many sources of pollution. Additionally, electricity generated largely from nuclear power plants will be increasingly substituted for other forms of energy having greater adverse environmental impact.

Regional Electrical Energy Growth

The 1970 Federal Power Commission estimate (fig. 5.5) of West Region future power requirements indicates a growth rate greater than that shown in the 1964 National Power Survey. This increased rate of growth follows a trend similar to that of the total electric energy growth in the contiguous United States. In 1965, the energy requirement of the West Region amounted to about 20 percent of the total for the nation. In 1990, the percentage is expected to be about 21 percent.

The 1970 Federal Power Commission study of the western region indicates that "total energy requirements for the West Region in 1965 were 212,550 megawatthours with a corresponding peak demand of 37,545 MW, and a load factor of 64.6 percent. For 1990. these are estimated at 1,232,800 megawatthours, 216,420 MW, and a load factor of 65 percent. Over this 25 year period this represents an energy growth of 5.8 times and an average compound rate of growth of 7.3 percent per year." [5]

Basically the same factors that influence national electrical growth will also affect regional growth; however, one key factor affecting national growth will have a more intensive effect upon some regional increases. As previously mentioned, the rate of population growth for the western region is substantially higher than that of the rest of the United States. For this reason, electrical energy increases in the West will tend to occur at a faster pace than for the nation.

ENERGY RESOURCES

National Energy Resources

Norman Brooks, professor of environmental science and civil engineering at California Institute of Technology, recently outlined the demand that total energy placed upon our natural resources (see fig. 5.6).

The total energy consumption in the United States in 1968 was 17.8 trillion kilowatthours (kWH equivalents)—equivalent to a continuous average use of 10,000 watts per person. In 1900 it was 2.2 trillion kilowatthours—an overall average growth in the intervening years of 3.1 percent per year (compounded). But in the period 1935 to 1968, the rate was 3.6 percent per year, and in the most recent four years, 1964–68, the rate of growth was 4.9 percent per year. A growth rate of 5 percent per year in the future would lead to an energy consumption of 76 trillion kilowatthours in the year 2000—over four times the present yearly usage (see fig. 5.7). If the growth rate were only 2 percent per year, the figure for 2000 A.D. would be less than half as much—34 trillion kilowatthours. The important point is that the growth rate, when compounded over many years, has an enormous leverage on how our resources and environment are going to be affected.

I am not a 'futurist' because I do not think we must plan for whatever the demands may be; on the contrary, it will be necessary for society to control its growth rate to keep energy use and its environmental effects within tolerable limits. The primary sources of energy were, for 1968:

Crude Petroleum	40.7%
Natural Gas	32.1%
Coal	21.9%
Natural Gas Liquids	3.8%
Hydro Electric Generation	1.3%
Nuclear Energy	0.2%
Total (17.8 × 10^{12} kWH)	100.0%

Fossil fuels accounted for 98.5 percent of the total, while the nuclear energy was only 0.2%. The nuclear fraction will grow rapidly in the future as it displaces fossil fuels in the production of electricity. Hydro-electric power is also a very small

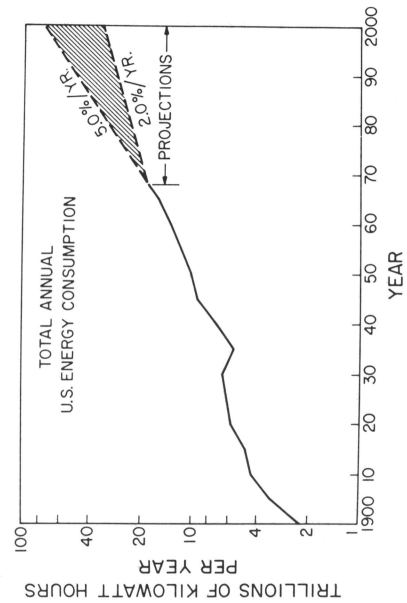

Fig. 5.6 Historic growth of United States annual energy consumption with a high and low projection to the year 2000.

111

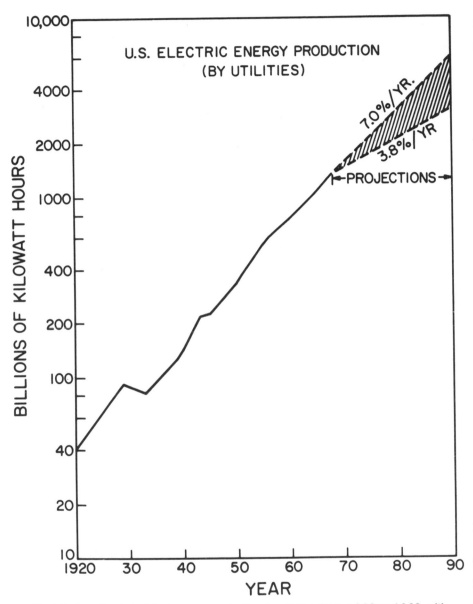

Fig. 5.7 Growth of electrical energy production by utilities from 1920 to 1968, with range of projections to 1990.

percentage and will probably continue to stay small because most of the feasible hydro sites have already been built. It should be noted that hydro power is the only energy source which uses the current energy budget of the earth rather than energy stored from some other geologic age. There are other small sources of energy—wood, refuse, geothermal heat—which do not usually appear in the statistical summaries, but the amount for the U.S. is probably only a few tenths of one percent of the total. Direct beneficial use of solar energy (agricultural, drying, heating, etc.) cannot be computed; the amounts mentioned here refer to only man-made energy distribution systems.[6]

Brooks' observation that "hydro power is the only [major] energy source which uses the current energy budget of the earth rather than energy stored from some other geologic age" is of interest. Energy sources which utilize the earth's current energy budget can be referrred to as income energy sources. Winds, tidal fluctuations, hydro power, and solar energy are all forms of income energy. If man were able to harness large amounts of income energy, pollution problems could be minimized, environmental effects of large dams and local thermal effects being the major remaining concerns. "Energy stored from some other geologic age," can be referred to as capital energy, that is, an energy investment given to us by nature which we may draw upon from time to time. Unfortunately, as man utilizes capital energy he is faced with the challenges of air pollution and both global and local thermal effects.

Mr. Brooks continues by discussing the resources needed to meet electrical power growth (see fig. 5.7).

Of the total energy consumed in 1968, only 8.0 percent—1.43 trillion kilowatt-hours—was converted to electric power. However, an additional 14 percent of the total was discharged as waste heat by the thermal power stations, operating at an average efficiency of 33 percent. Hence 22 percent of the total energy supply was used to run the electric power systems.

Electric power production by utilities was 1.33 trillion kilowatthours—a growth of 9 percent over the preceding year. (The small remainder—0.10 trillion kilowatt-hours—was produced by industry for in-plant use.) For the period 1920–68, the average growth rate was 7.5 percent per year—or doubling every 9.5 years. If the demand continues to grow at 7 percent per year, the production by utilities will have to increase over four-fold by 1990 to 5.9 trillion kWH. Even a much lower projection of only 3.8 percent growth per year leads to 3.0 trillion kWH. The difference between these two production projections for 1990 is more than twice the current production, which illustrates the enormous impact that the growth rate has on problems of power plant siting.[6]

Regional Energy Resources

As evidenced by figure 5.5, hydro-electric generation, which originally satisfied most of the aggregate electrical generation requirements of the West Region and is projected to continue to increase in absolute amounts throughout the forecast period, is declining in its relative share of generation, decreasing from 78 percent in 1952 to 57 percent in 1965. Thermal generation is projected to exceed hydro generation by 1975 and to satisfy an ever increasing share thereafter. Oil and gas fired thermal generation, representing from about 85 to 95 percent of thermal generation from 1950 to 1965, is expected to increase in absolute amounts until about 1975 and to decline thereafter—its generation requirement being eroded by more economic base-loaded coal and nuclear generation. Coal and nuclear generation sources are projected to satisfy the bulk of the increase in generation requirements after 1970.[5]

Increasing amounts of thermal energy forms will be utilized to provide electrical energy. The relative roles of income energy sources for electrical generation will continue to decline until such time as man develops ways to harness large income energy sources, such as solar power.

CHALLENGE TO MANKIND

Population must be controlled and stabilized by man, for the alternative is that population will be stabilized or reduced by natural factors beyond man's control. If man reduces any one of the key life support resources, such as water or air, to a nonusable form by neglect or design, the safe level of population will drop.

The hallmark of past decades has been growth—often growth for growth's sake. The growth ethic is embodied in both religious code and mores. The present state of affairs warrants a shift from the traditional growth ethic to a quality and conservation oriented society. Some electrical utilities have recognized this need and have begun to discontinue "promotional advertising"; for example, recent ads by more than one utility in the West urge people to conserve energy and suggest ways by which the homeowner may do so.

Industry must continue to develop the technology to clean up manufacturing processes. Fossil-fuel power plants are major pollutant sources but, as is evidenced in earlier portions of this paper, fossil fuels will be required to provide major portions of our total energy needs for many years in the future. In view of the future increased use of fossil fuels, technology must be developed to first control and later eliminate the pollutants associated with fossil-fuel combustion.

In general, the use of income energy tends to minimize pollution problems. It is hoped that efforts will continue to develop the technology necessary to harness additional sources of income energy.

Recognizing that the bulk of our energy needs will be consumed in the urban environment, what means of satisfying these requirements will produce the minimum adverse impact on the environment?

The major environmental impacts of the production, transportation, and utilization of energy are:

1. Air pollution:	SO_x, NO_x, and particulates
2. Thermal effects:	increased air temperature, increased atmospheric humidity, and/or increased cooling water temperatures
3. Presence:	space and land use and aesthetic problems

The challenge facing electric utilities is awesome. In response to a growing economy and population, I believe 350 GW of electrical capacity must be added to the United States by 1980, and some 540 GW more by 1990. These estimates are based on a belief that the rate of increase in electrical use will decline due to changing social values, improved thermal efficiencies, a growing saturation of energy uses, and other factors.

The large portions of manpower and capital needed to meet this national demand is a challenge in itself, but an even greater task is the necessity to meet the challenge with a minimum adverse impact upon our environment.

Reference notes

1. Fremont Felix, "Annual Growth Rate on Downward Trend," *Electrical World,* 6 July 1970, p. 30.
2. Hans H. Landsberg, *Energy in the United States* (New York: Random House, 1965), pp. 59–60.
3. *United Nations 1968 Statistical Yearbook,* Statistical Office of the United Nations, Department of Economic and Social Affairs (New York, 1969).
4. "Population Heads for a Zero Growth Rate," *Business Week,* 24 October 1970, p. 102.
5. Federal Power Commission, "The 1970 National Power Survey," pt. 3 (United States Government Printing Office, 1970).
6. Norman Brooks, "Energy and the Environment," *Engineering and Science* (California Institute of Technology, January 1971), p. 21.

6. Transportation Energy Requirements

R. C. Amero
Gulf Research and Development Company

In the past thirty years, liquid hydrocarbon fuels have almost completely displaced coal, wood, and other sources of energy for transportation. There is little doubt that the market will continue to expand, and liquid fuels (derived from oil shale, tar sands, and coal in addition to petroleum) will continue to dominate.

The developments in fuels and engines are closely coordinated and interdependent. To project the next decade in fuels for transportation it is helpful to start with an overview of today's engines, fuels, and refining techniques, and then project the trends in each of these areas.

ENGINES AND FUELS PRESENTLY USED FOR TRANSPORTATION

Engines

Gasoline engines obviously dominate the market for units below 400 HP, especially for intermittent use, such as automobiles, outboard engines, motorcycles, and private aircraft.

Diesels predominate where higher output and more continuous duty are involved, as in tractor-trailers, buses, locomotives, and commercial marine vessels. Diesels are also being promoted with some success in automotive sizes, especially for taxicabs, delivery trucks, and pleasure boats. Diesels should be considered in three separate classes:

117

1. High speed: the familiar trunk-piston type engines operating over 1200 rpm and requiring kerosene or refined distillate fuel;
2. Intermediate speed: locomotive and marine engines, 300 to 1200 rpm, which operate on distillate fuel at low maintenance cost but are sometimes operated on blends containing residual fuel;
3. Low speed: prime movers for deep-draft vessels which operate at 300 rpm or less—the slow speed allows combustion time for very heavy fuels which are usually blends of Bunker C with 5 to 50 percent distillate diesel fuel.

Steam power plants for transportation are now limited primarily to large ships, and their virtues as compared with low speed diesels are debated strongly in marine circles. They burn the cheapest of the heavy, tarry fuel oils.

Gas turbines now dominate the commercial aircraft market and are being promoted vigorously in all prime mover markets. For land and sea transportation, their penetration in the market has been small; their greatest success has been in the higher horsepower, heavy duty applications such as naval and merchant marine, large trucks, and locomotives.

Gas turbines fall into two distinct classes with different basic fuel tolerances:

1. The jet aircraft turbine, weighing 0.5 to 1.0 lb. per HP, is put to work on land or sea with little modification. Usually a free turbine is added to generate shaft horsepower instead of thrust from the exhaust gases. These turbines are critical of the thermal stability and combustion properties of a fuel, and require a refined kerosene or distillate. Automotive turbines are also compact and lightweight but usually have a heat recovery device for better part-load economy. Specific weight is 2 lb. or more per HP.
2. "Industrial" gas turbines use cast housings, larger combustion chambers, and more conservative ratings. They weigh 4 to 7 lb. per HP and operate well on fuels that as yet cannot be digested by the aircraft gas turbines, but as they become more efficient they become more critical of fuel purity.

Pipeline movements represent a separate category of transportation where pumping energy will most generally be taken from the fuel flowing through the line. Spark-ignition natural gas engines are dominant on natural gas lines and diesels on liquid product lines, but gas turbines are competing with both. An automatic distillation unit has been developed which supplies clean pumping fuel from the crude in a crude oil line.[1]

Fuel Consumption in Perspective

Figure 6.1 and table 6.1 show that transportation consumes one-quarter of the energy used in the United States and gasoline, diesel (distillate) and jet fuel account for 80 percent of the transportation energy. More than 95 percent of the horsepower in the United States is installed in transportation equipment (table 6.2). Figure 6.2 shows the dominance of gasoline in the field of transportation fuels.

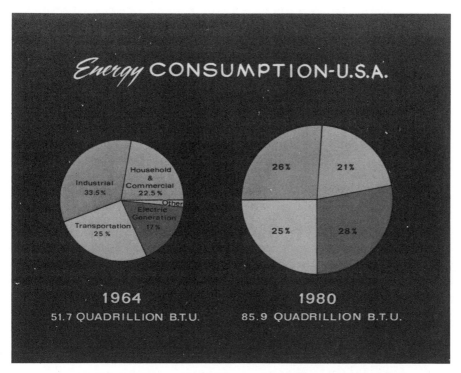

Fig. 6.1 The changing profile of energy consumption expected by 1980.

There are over 105 million vehicles in the United States, or half of the world total (table 6.3), and they consumed over 5.5 million bbl. of fuel per day in 1969 out of the United States total of 14.2 million bbl. of petroleum products per day.[2] United States motorists drove the equivalent of more than 40 million trips around the world in 1968. Outside of the United States, gasoline demand (2.5 million bbl./day) is a smaller fraction of the total energy requirement, as shown in table 6.4.

TABLE 6.1
United States Consumption of Energy
(trillion Btu.)
1968 * and 1980 †

	Household and Commercial		Industrial		Transportation		Electricity Generation		Miscellaneous	Total	
	1968	1980	1968	1980	1968	1980	1968	1980	1968	1968	1980
Anthracite	121	50	80	50	—	—	56	150	1	258	250
Bituminous and lignite	447	200	5,536	3,476	12	—	7,074	12,416	—	13,069	16,092
Natural gas	6,451	10,367	9,258	11,538	610	693	3,245	3,918	—	19,564	26,516
Petroleum	(6,581) †	7,362 †	(4,474) †	7,168 †	(14,513)	20,650	1,181	862	(295)	(27,044) †	36,042 †
LPG	686	—	83	—	120	—	—	—	14	904	—
Military jet fuel	—	—	—	—	679	—	—	—	—	679	—
Commercial jet fuel	—	—	—	—	1,260	—	—	—	—	1,260	—
Gasoline	—	—	—	—	10,265	—	—	—	—	10,265	—
Kerosene	446	—	133	—	—	—	—	—	5	585	—
Distillate	3,233	—	357	—	1,240	—	18	—	178	5,025	—
Residual	1,232	—	1,075	—	804	—	1,163	—	—	4,275	—
Still gas	—	—	899	—	—	—	—	—	—	899	—
Coke	—	—	327	—	—	—	—	—	—	327	—
Nonfuel uses *	984	—	—	—	—	—	—	—	—	984	—
Miscellaneous	—	—	—	—	—	—	—	—	99	99	—
Hydropower	—	—	—	—	—	—	2,359	2,675	—	2,359	2,675
Nuclear	—	—	—	—	—	—	130	4,361	—	130	4,361
Total	13,600	17,979	19,348	22,231	15,135	21,343	14,045	24,382	296	62,424	85,934

SOURCES: 1. *Minerals Yearbook*, 1968.
2. W. A. Vogely, "Pattern of Energy Consumption in the United States," American Chemical Society, Division of Fuel Chemical Reprints 9, no. 2 (1965): 205–21.

* Lubes, wax, coke, asphalt, road oil, petrochemical feedstock
† Excluding nonfuel uses

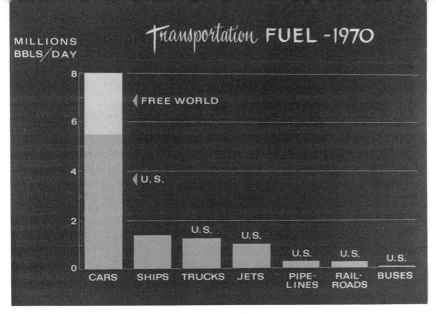

Fig. 6.2 Relative fuel requirements of various transportation modes.

TABLE 6.2
Total Horsepower of All Prime Movers, 1940 to 1969

Item	1940	1950	1955	1960	1965	1968	1969 (preliminary)
Total Horsepower	2,773	4,868	7,158	11,008	15,096	17,828	18,781
Automotive *†	2,511	4,404	6,632	10,367	14,306	16,854	17,742
Nonautomotive	262	464	526	641	790	974	1,039
Factories	22	33	36	42	48	52	53
Mines	7	22	31	35	40	43	44
Railroads ‡	92	111	60	47	44	58	53
Merchant ships and sailing vessels	9 §	23 §	24 §	24	24	20	19
Farms ‖	70	165	212	240	272	292	302
Electric central stations †	54	88	138	217	307	372	404
Aircraft ‡#	7 §	22 §	26 §	37	55	137	164

SOURCE: United States Bureau of the Census, "Statistical Abstract of the United States: 1970," 91st ed. (Washington, D.C., 1970).
NOTE: In millions. As of January, except as noted. Prior to 1960, excludes Alaska and Hawaii, except as noted. Prime movers are mechanical engines and turbines, and work animals, which originally convert fuels or force (as wind or falling water) into work and power. Electric motors, which obtain their power from prime movers, are excluded to avoid duplication.
* Includes passenger cars, trucks, buses, and motorcycles
† As of July 1
‡ Beginning 1965, not strictly comparable with earlier years
§ Includes Alaska and Hawaii
‖ Includes about 1.3 million HP in work animals and 22,000 in windmills
Includes private planes and commercial airliners

TABLE 6.3
Motor Vehicle Registrations
(cars, trucks, buses: in thousands)

	1950	1955	1960	1965	1966	1967	1968	1969 (preliminary)
World	70,424	95,896	126,908	177,995	190,306	205,949	216,608	(NA)
United States	49,161	62,689	73,869	90,360	94,193	99,958	101,039	104,702
Other North and Central America	3,154	4,942	6,759	8,235	9,160	9,517	10,169	(NA)
South America	1,270	1,835	2,995	5,103	5,573	6,122	6,644	(NA)
Europe	12,529	19,611	35,751	56,483	61,097	66,977	72,191	(NA)
Africa	1,241	1,829	2,450	3,394	3,509	3,759	4,029	(NA)
Asia	1,221	2,212	3,556	9,739	11,817	14,477	17,175	(NA)
Oceania	1,848	2,778	3,528	4,681	4,957	5,138	5,361	(NA)

SOURCE: United States Bureau of the Census, "Statistical Abstract of the United States: 1970," 91st ed. (Washington, D.C., 1970).

TABLE 6.4
World Consumption of Energy
(United Nations data for 1968)

	Total Energy Consumed			Gasoline Consumption in Each Area	
	Million Metric Tons of Coal Equivalent	Percentage of World Consumption	Kilograms of Coal Equivalent Per Capita	Percentage of All Energy Used	Percentage of All Liquid Fuels Consumed
World	6,016	100	1,727	11:6	28.9
North America	2,225	37.6	10,155	18.2	44.3
United States	2,078	34.6	10,331	18.5	45.5
Canada	176	2.9	8,480	15.4	32.1
Caribbean America	120	2.0	1,032	21.2	31.5
Central and South America	98	1.6	653	19.9	26.0
Western Europe	1,161	19.3	3,312	8.3	16.0
United Kingdom	277	4.6	5,004	7.1	18.6
Western Asia	56	0.9	575	10.7	14.8
Far East	436	7.3	402	7.2	13.0
Oceania	71	1.2	3,864	16.9	36.4
Africa	99	1.6	294	10.6	26.9
Other Countries	1,719	28.6	1,553	4.8	22.2
USSR	965	16.0	4,058	—	—

SOURCE: "World Energy Supplies 1965–68," Statistical Papers Series J no. 13, Department of Economic and Social Affairs, Statistical Office of the United Nations (New York, 1970).

To satisfy the market, refineries have been increasing the yield and octane number of motor gasoline for fifty years. Meanwhile, distillate demand has grown steadily with the advent of automatic home heat in the 1940s, dieselization of trucks and railroads in the 1950s, and "turbinization" of aircraft in the 1960s. Consumption of jet aircraft fuel passed the million barrel per day level in the United States in 1969 and is headed to the 1.5 million barrel level by 1975.[3,4] Thus, there is a large gasoline demand linked directly to the automotive population, and a smaller but faster growing distillate demand.

Gasoline is unique in its requirement for "octane number" to satisfy spark-ignition engines. On the other hand, distillate fuel for gas turbines, diesel engines, and furnaces are similar to each other in quality requirements, such as clean-burning characteristics, storage stability, and stability against sludging or lacquer formation when exposed to hot sections in the equipment. Quality controls are less stringent on residuals although sulfur, stability, and compatibility limits are imposed for some uses.

In summary, petroleum energy is marketed in three sectors: (1) gasoline for automotive, small aircraft, and miscellaneous small engines; (2) distillates for burners and medium to large engines; and (3) residuals and blends for medium and large burners and large diesel engines. Fuels for transportation are similar to, and competitive with, petroleum fuels used for space heating, industrial processing, and electrical generation.

REFINING TECHNIQUES AND TRENDS

The simplest refinery fractionates crude into naphtha, middle distillate (diesel fuel), and residue (Bunker C or No. 6 Fuel Oil). This is a fortuitous combination for emerging nations which need industrial fuel; they can upgrade the naphtha into the small volume of motor gasoline needed for their economy. Typical petroleum fractions from atmospheric distillation are shown in figure 6.3.

The next step in refinery sophistication is vacuum distillation so that higher-boiling fractions, formerly left in residual fuel, can be recovered separately. This heavy distillate is refined into lubricating oils or cracked to gasoline. Heavy distillate contains high molecular weight compounds with aromatic and naphthenic rings linked or fused together; cracking breaks naphthenic rings, leaving fragments that are low in molecular weight and more aromatic than the original distillate. Even though catalysts and special reactors are used to control the cracking, only half of the heavy distillate is converted into gasoline. The remainder is a spectrum of hydrocarbon fragments from gas to tar, high in aromatic content, which must be fractionated and blended into other refinery products.

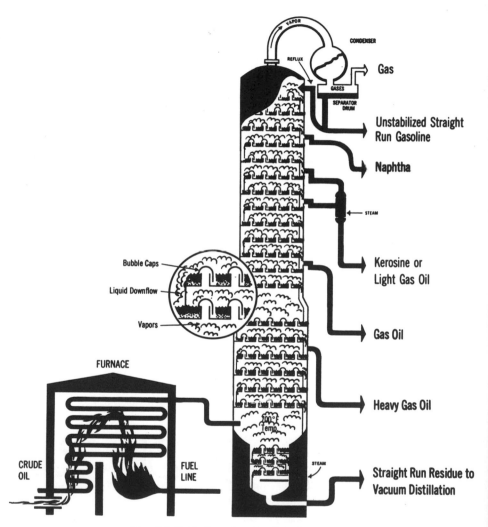

Fig. 6.3 Typical modern fractionating tower.
SOURCE: American Petroleum Institute.

124

Aromatics have good octane ratings but they depreciate the cetane number of diesel fuel and the clean-burning qualities of gas turbine and furnace fuels. Thus, cracking increases the gasoline yield and, to a point, improves its quality; but cracking is not beneficial to other petroleum fuels.

Residual fuel might represent 50 percent of the refinery yield from atmospheric distillation. With vacuum distillation followed by "visbreaking" of the residual (heating to crack out a little more distillate), residual yield has been reduced to 6 percent of the total crude refined in the United States. In fact, coking of residual "to extinction" is economically desirable in the United States except for the limited market for by-product coke.

The octane race forced refiners to add an array of upgrading treatments. Alkylation converts refinery gases into high octane liquids. Isomerization of butanes and pentanes increases the octane numbers of these blending components. Reforming of naphtha has been widely adopted to increase the aromatics, with a resultant increase in octane number. Reforming releases hydrogen as a by-product, which triggered the next refining development.

With free hydrogen available, the refiner had a tool for restoring diesel and home heating fuel to a higher quality level. Hydrogenation gave better color, lower sulfur content, and better stability to fuels containing cracked distillate. Hydrogenation proved uneconomical in 1936, but in the mid-

TABLE 6.5
Crude Oil and Major Processing Capacity
(thousands of barrels per day, as of 15 January 1970)

Country	Number of Plants	Crude Capacity	Vacuum Distillation	Thermal Processes	Catalytic Cracking	Catalytic Reforming	Hydrogen Processes
North America	310	14,398	4,841	1,918	4,761	2,983	4,836
Canada	41	1,303	493	81	354	219	545
Mexico	6	501	228	59	107	85	160
United States	263	12,595	4,120	1,778	4,300	2,679	4,131
Central America and Caribbean	22	1,766	221	484	169	107	201
South America	60	2,533	473	485	300	79	41
Western Europe	168	13,068	1,386	804	700	1,627	2,400
United Kingdom	22	2,028	344	81	151	257	272
Africa	29	709	33	31	18	108	92
Middle East	29	2,229	303	147	86	134	331
Asia and Far East	87	5,376	828	186	364	509	1,076
Free World total	705	40,080	8,084	4,054	6,397	5,548	8,977
USSR and Soviet Bloc	NA	6,437	—	—	—	—	—

SOURCE: World Petroleum Report 16 (1970): 28–29.

TABLE 6.6
Variation in Fuel Properties With Refining

	Typical Diesel Fuel *				Typical Residuals		
	Straight Run	Straight Run Plus Cracked		Hydro-cracked	Topped Venezuelan Crude	Visbroken Texas Residual	Topped Low-Sulfur African Crude
		Without Hydro-genation	With Hydro-genation				
Gravity: °API	39.1	35.6	35.0	41.7	14.4	10.1	20.4
Sp. gr., 60°/60° F.	0.8294	0.8468	0.8499	0.8170	0.9698	0.9993	0.9315
Pour point: ° F.	−20	−15	−10	−10	+20	+30	+80
Sulfur: % by wt.	0.04	0.22	0.12	0.05	2.4	2.2	0.54
Carbon-hydrogen ratio	—	—	—	—	7.86	8.91	7.31
Hydrocarbon analysis % by volume							
Paraffins	36.3	69.0	35.4	48.5	26.7	—	—
Cycloparaffins	42.7		29.7	34.4		—	—
Olefins	1.0	2.0	2.0	1.5	—	—	—
Aromatics	20.0	29.0	32.9	15.6	45.4	—	—
Cetane index	51.8	47.0	48.0	57.0	—	—	—
Distillation: ° F.							
Initial boiling pt.	376	353	376	358	—	—	—
10%	410	419	430	410	—	—	—
50%	478	493	506	478	—	—	—
90%	570	555	586	572	—	—	—
Final boiling pt.	610	593	636	626	—	—	—

* From Sweet Midcontinent and Gulf Coast crudes, USA.

1950s, free hydrogen improved catalysts and the problems with cracked distillates made such plants feasible.[5]

Hydrogen has been so useful that refiners are beginning to use more drastic hydrogenation, even where hydrogen is no longer free. Hydrocracking consumes 1000 to 4000 standard cu. ft. (SCF) of hydrogen per bbl. of product and converts aromatics into naphthenes, paraffins, and isoparaffins. For comparison, the milder hydrogenation processes consume 50 to 600 SCF/bbl. to remove the more active sulfur and nitrogen compounds, decarboxylate naphthenic acids, and chemically saturate some olefins. The operating pressures of 1000 to 1500 psi for hydrocracking require more hydrogen and much higher investment in reactor and piping.

Table 6.5 shows the crude oil distillation capacity in various parts of the world in 1970 plus the "downstream processes" for converting crude fractions into fuels. Table 6.6 shows some variations in distillate and residual fuels related to different refining treatments.

Tar Sands, Shale Oil, and Coal

Hydrogen technology is the key for making liquid fuels of good quality from raw materials other than crude oil, such as tar sands, shale oil, and coal. Tar sands are viscous petroleum mixed with sand. After mining and sand separation, the crude can be refined in a conventional manner. Shale oil is a rock that is mined and retorted to yield a liquid containing hydrogen, carbon, and sulfur, plus a few percent of nitrogen and oxygen. Retorting of coal has been practiced for years but hydrogenation for upgrading the liquid products is new in the technology. Hydrogenation is the common denominator for making marketable products from tar sands and shale oil as well as from coal. Properties of crudes produced by hydrogenation of these starting materials are shown in table 6.7. The tar sand and shale oil derivatives, after hydrogenation, could be handled in conventional refineries. The coal derivatives require hydrocracking to be equal in quality to conventional petroleum products.

Oil imports to the United States are limited, to encourage a healthy domestic petroleum industry. Depending on import restrictions, and the development of domestic reserves such as the North Slope of Alaska,[6] the demand for domestic crude will force the development of alternate raw materials for liquid fuels. The First National City Bank of New York has forecast a million barrels of synthetic fuel per day by 1980, half from coal in the United States and half from shale and oil sands from the United States and Canada. Great Canadian's new facilities at Fort McMurray, Alberta, produced 27,400 bbl./day from oil sands in 1969, proving that the operation is technically feasible, if not yet profitable. The Oil Shale Corporation, Cleveland Cliffs, Ohio, program is pointed towards the production of 50,000

TABLE 6.7
Typical Nonpetroleum Fuel Sources

Source	Oil Shale		Tar Sands		Coal	
	Raw Crude	Hydro-genated Crude	Raw Crude	Hydro-genated Crude	Raw Crude	Hydro-genated Crude
API gravity	21.2	41.4	25.4	31.5	1.3	21.5
Density: lb./gal., 60°/60° F.	7.717	6.814	7.510	7.228	8.875	7.702
Ultimate analysis:						
Carbon: %	84.6	86.1	85.35	87.10	83.7	87.20
Hydrogen: %	10.84	13.84	11.69	12.69	8.6	11.48
Oxygen: %	2.0	0.12	0.34	0.04	6.5	0.72
Nitrogen: %	1.6	0.01	0.24	0.07	1.0	0.25
Sulfur: %	0.77	0.02	2.38	0.10	0.2	0.005
Ash: %	—	—	0.025	—	—	—

bbl./day of oil from shale in the next few years in the Picaence Creek Basin, Colorado.[7] Consolidated Coal Company's Project Gasoline and Hydrocarbon Research, Inc.'s Project H-Coal are similar in that both dissolve or suspend coal in a hydrocarbon solvent before hydrogenation. On the other hand, FMC Corporation's Project COED (Char Oil Energy Development) involves coal pyrolysis followed by hydrogenation of the condensed vapors.[8] Development continues on these processes toward the time when economics will be favorable.

Oxygen is another refinery tool but it has been used almost exclusively for the manufacture of chemicals, such as alcohols, glycols, and organic acids. Oxygenated hydrocarbons contain less energy than pure hydrocarbons and therefore have a built-in penalty with respect to transportation costs and performance. Long-chain oxygenated compounds, particularly certain esters, represent a balance between clean-burning and high energy characteristics that may be of interest as air pollution controls are tightened.

In summary, petroleum refining has developed in the direction of more and better gasoline, but technology is available to tailor the yield and quality of other liquid fuels to changing market demands. Technology is in an advanced pilot plant stage for the use of coal and oil shale.

FUTURE FUELS AND ENGINES FOR TRANSPORTATION

There is universal agreement that energy requirements will continue to increase. Table 6.1 shows a prediction that transportation will account for a quarter of United States energy consumption in 1980, as it does at present, and petroleum will fuel 95 percent of the transportation.

The recent study by P. A. Laurent of energy demands for Western Europe predicts a more rapid increase in transportation requirements than in other sectors of the economy.[9] He foresees a 1980 demand for 247 million tons of petroleum fuels for transportation, representing 30 percent of the petroleum energy and 17.3 percent of all energy in the area. The analysis by S. F. Jefferson and J. G. Trimmer in 1965 showed that transportation was consuming 19 percent of all energy used in Western Europe and about one-third of the petroleum.[10]

Automotive: Gasoline

There were over 75 million automobiles in the United States in 1967 and over 90 million in 1969, with predictions for 120 million by 1980 and 240 million by 2000.[11] Nineteen separate surveys predict that the United States car population of 1965 will be doubled between 1982 and 1993.[12] Table 6.3 shows that the rate of increase in vehicle registration is proportionately high all over the world. Because of the weight-cost-performance advantages of the gasoline engine, typical pronouncements by automobile manufacturers predict that the population of gasoline-fueled automobiles will continue to increase "as far ahead as we can see." [13] The consumption of gasoline in the United States expands each year by about 200,000 bbl. per day.

Figure 6.4 shows that Research Octane Number has leveled off near 100 for premium and 94 for regular gasoline in the United States. Motor Octane Number continued to rise because of the increasing use of alkylate (isoparaffinic) to supplement the cracked and reformed gasolines (aromatic). Both octane scales have a bearing on engine knock, so that the improved Motor Octane Numbers indicate continuing modest improvement in antiknock performance on the road. .

Automotive: Pollution

The octane race of the 1950s was superseded by air pollution questions in the 1960s. The National Center for Air Pollution Control stated that increased emissions in a metropolitan area will be due mainly to the spreading of high-traffic density areas rather than to increased traffic density in the core area.[14] Their studies in eight urban areas indicated that drastic reductions would be required to reduce the emissions in 2010 A.D. below 50 percent of the 1967 level. As compared with 1967 models:

1. Emissions on 1968 to 1972 automobiles should be reduced 50 percent.
2. Emissions on 1973 to 1977 automobiles should be reduced 80 percent.
3. Emissions on later automobiles should be reduced 95 percent.

Table 6.8 shows that transportation is responsible for 42 percent of United States air pollution, overwhelmingly in the form of carbon monox-

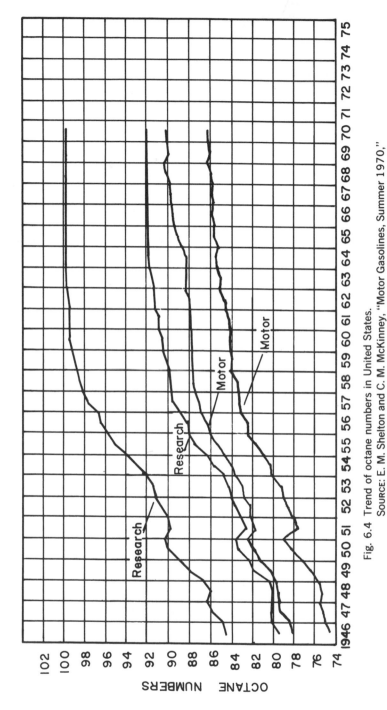

Fig. 6.4 Trend of octane numbers in United States.
SOURCE: E. M. Shelton and C. M. McKinney, "Motor Gasolines, Summer 1970," *Petroleum Products Survey No. 68*, Mineral Industries Survey, United States Department of the Interior, Bureau of Mines, Washington, D.C.

130

TABLE 6.8
Air Pollutant Emissions, 1968
(quantity in millions of tons per year—estimates)

Source Category	Total	Carbon Mon-oxide	Sulfur Oxides	Hydro-carbons	Partic-ulates	Nitrogen Oxides
Quantity						
Total	213.8	100.1	32.8	32.0	28.3	20.6
Transportation	90.5	63.8	0.8	16.6	1.2	8.1
Fuel combustion						
(stationary)	45.5	1.9	24.0	0.7	8.9	10.0
Industrial processes	29.3	9.7	7.3	4.6	7.5	0.2
Refuse disposal	11.2	7.8	0.1	1.6	1.1	0.6
Miscellaneous	37.3	16.9	0.6	8.5	9.6	1.7

SOURCE: United States Bureau of the Census, "Statistical Abstract of the United States: 1970," 91st ed. (Washington, D.C., 1970).

ide. The announced restrictions on automotive emissions by California and the federal government are shown in table 6.9.

Changes in gasoline composition do not always produce the desirable changes expected in exhaust emissions. Reduction in gasoline volatility to control evaporative emissions often has an adverse effect on engine operation and increases the exhaust emissions. Limitations on aromatics and olefins in gasoline have minor influence on exhaust composition because these compounds are also created by combustion conditions in the engine, as are the nitrogen oxides.

Automobile manufacturers reduce engine emissions each year but a "smog-free" car has yet to be developed. This is one which the California Air Resources Board has defined as emitting less than 50 ppm. hydrocarbons, 0.5 percent carbon monoxide, and 100 ppm. nitrogen oxide after 50,000 miles. Blowby fumes from crankcases are being recirculated to the air intake and carburetor improvements reduce exhaust emissions. Starting with 1970 models, all cars for sale in California have devices to trap evaporative emissions in activated charcoal or in the crankcase after engine shutdown, in which case the emissions are recovered when the engine is started again. Nationwide, all 1971 models are equipped with evaporative emission controls.

Low-lead and no-lead gasolines have appeared in response to the claim that automobiles cannot meet the 1975 limits on particulate matter emissions with leaded gasoline and that lead poisons many catalysts of potential future emission control devices. Also, lead compounds plug the orifices in control devices developed for recirculating 5 to 20 percent of the exhaust

TABLE 6.9
Legislative Restrictions on Automotive Emissions
California Vehicle Emissions Standards A.B. 357 and Federal Standards *

Car model year	1970	1971	1972	1973	1974	1975
Light duty ‡						
Hydrocarbons: gm/mile (ppm.†)						
California	2.2(180)	—	1.5(125)	—	—	—
United States	2.2(180)	—	—	—	—	0.46
Carbon monoxide: gm/mile (vol. %†)						
California	23(1.0)	—	—	—	—	—
United States	23(1.0)	—	—	—	—	4.7
Nitrogen oxides: gm/mile (ppm.†)						
California	—	4(1075)	3(800)	—	1.3(350)	—
United States	—	—	—	3	—	3.0
Evaporative emissions: gm/test						
California	6	—	—	—	—	—
United States	—	6	—	—	—	—
Heavy duty, California §						
Hydrocarbons: ppm.	275	—	180	—	—	—
Carbon monoxide: percentage	1.5	—	1.0	—	—	—

* Exhaust analyses, determined and expressed in "ppm." or "vol. %," are converted to "gm/mile" from standardized charts for twelve different weight classes of vehicles. In California the limits apply to each car; federal standards apply to the average car.
† Approximate concentration in the exhaust of a "typical" vehicle of 3500–4000 lb. with automatic transmission.
‡ Vehicles with gross weight less than 6001 lb. and having engines greater than 50 cu. in. displacement.
§ Vehicles with gross weight greater than 6000 lb.

gas back through an engine to reduce NO_x emissions. Many 1971 model automobiles have reduced compression ratios and are designed to operate on 91 Research octane number gasoline; however, public acceptance has been slow because low-lead and no-lead gasolines are generally more costly than leaded gasolines of the same octane number. The whole situation is fluid and further changes in gasoline can be expected in the future.

Automotive: Alternate Fuels for Spark-Ignition Engines

Limited use is already being made of both liquefied and compressed natural gas in vehicles operating in Southern California, with very favorable results reported.

Propane has been a motor fuel, as the principal component of LPG, for thirty years. LPG (mixtures of butane and propane held in the liquid state at 80 to 150 lb. per square inch) offers the advantage of reduced engine maintenance and significantly reduced exhaust emissions as compared with gasoline burned in the same engine. A typical test with four police cars in Chandler, Arizona, showed substantially lower operating costs with propane.[15]

Savings with propane derive primarily from the lower fuel price, and this situation would change if a substantial demand developed for propane. The United States uses five and one-half times as much gasoline as LPG; in transportation alone, the gasoline:LPG ratio is about 60:1. However, LPG is destined for increased use in fleets which (1) idle for long periods, (2) operate in high pollution areas, and (3) operate over a small area so they can be refueled at a central point. Chicago has operated 1500 city buses on LPG for the past decade. Almost a thousand buses in Vienna use this fuel. Route delivery trucks, police patrol cars, and taxis are also good prospective users.

Alcohols receive sporadic attention as fuels, especially as a means for utilizing surplus products of agriculture. Not long ago, additional interest developed from the possibility that use of ethyl alcohol might result in lowered hydrocarbon emissions.[16] Basic fuel properties of methyl and ethyl alcohol are compared in table 6.10 with ammonia, hydrogen, benzene, isooctane, methane, propane, and reformed hexane.[17] From an overall view of physical properties, heat of combustion, and typical cost, it is difficult to compete with petroleum hydrocarbons.

Automotive: New Propulsion Systems

Several automotive power plant systems are compared in table 6.11. Of the fuel-burning engines, gas turbines generally offer the lowest emissions. Most significant of the current automotive projects are the Ford 707 and GM GT 309 turbine engines for truck and bus application. Gas turbines compete more directly with diesel engines (which are also relatively low in emissions and are more competitive in larger sizes) than with gasoline engines.

Several years hence, there might be a market for a small, electric shopper–type vehicle in small horsepower sizes. However, table 6.12 shows that the energy and power from the best batteries barely approach the minima from a gasoline engine.

Tables 6.11 and 6.12 do not include fuel cells, which could replace batteries and never need recharging. Fuel cells still require exotic fuels such as hydrogen, hydrazine, or methanol or they require an expensive (platinum) catalytic surface or a unit to steam-reform kerosene into a hydrogen-rich gas. Hydrogen-oxygen cells currently show an energy density of about 13 watt-hr. per lb. There is much incentive to develop fuel cells to accept a liquid primary fuel, such as kerosene, for automotive uses.[18]

TABLE 6.10
Fuel Properties

Fuel	Chemical Formula	Molecular Weight	C/H Ratio	Specific Gravity (60° F.)	Heat of Formation Kcal/gm-mole	Heat of Combustion Btu./lb.	Boiling Point C.°	Freezing Point C.°
Methane	CH_4	16	0.25	0.415 ‡ / 0.0407 *	-17.85	21,480	-161.49	-182.48
Propane	C_3H_8	44	0.375	0.5853 § / 0.112 *	-24.75	19,916	-44.5	-189.9
Isooctane	C_8H_{18}	114	0.445	0.702	-49.70	19,080	125	-56.5
Benzene	C_6H_6	78	1.00	0.884	19.7	17,270	80.1	5.5
Reformed hexane	†	(17.1)	(0.233)	0.0433 *	(-31.31)	(9,200)	gas mixture	gas mixture
Methyl alcohol	CH_3OH	32	0.25	0.796	-47.90	8,580	64.96	-97.8
Ethyl alcohol	C_2H_5OH	46	0.33	0.794	-56.0	11,550	78.5	-117.3
Ammonia	NH_3	17	0	0.71 ‖ / 0.0433 *	-11.04	8,000	-33.35	-77.7
Hydrogen	H_2	2	0	0.0051 *	0	51,608	-252.5	-259.14

SOURCE: E. S. Starkman, "Alternative Fuels for Control of Engine Emissions" (Paper 69-222, Air Pollution Control Association 62nd Annual Meeting, New York).

‡ Liquid methane at $-164°C.$, 1 atm
§ Liquid propane at $-45°C.$, 1 atm
‖ Liquid ammonia at $-32°C.$, 1 atm
* Density of gas at STP (1 atm, 25°C.)-lb/ft³
† CH_4 — 35.3%
  H_2 — 30.0%
  CO — 3.3%
  CO_2 — 16.4%
  H_2O — 15.0%

TABLE 6.11
Comparison of Automobile Power Plants

	Emissions	Initial Cost	Fuel Cost	Maintenance Cost	Duty Matching	Performance	Durability	Weight	Space	Fuel Range and Convenience	Noise, Vibration	State of Development	COMMENTS
Gasoline, piston	P	G	G	M	M	G	M	G	G	G	M	G	Present motive power.
Low-emission piston	M	G	G	M	M	G	M	G	G	G	M	M	Present motive power.
Gas turbine	M	M	M	G	G	G	G	G	M	G	M	G	Expensive alloys; high tooling cost; very low emissions. May appear soon in trucks and prestige cars.
Fuel injection	M	M	G	M	M	G	M	G	M	G	M	M	Potential low emissions, but difficult to achieve with production tolerances and cost limits.
Stratified charge	M	M	G	M	M	G	M	G	G	G	M	M	Potentially lower emissions but fuel injection adds to cost.
Diesel	M	M	G	M	M	G	M	M	G	G	M	G	Low emissions of hydrocarbons and CO but costlier, noisier, heavier, more vibration-prone than gasoline engines.
Wankel	P	M	G	M	M	G	M	G	G	G	M	M	Tests indicate higher emissions from this rotary engine than from piston engines.
Free-piston, turbine	M	M	M	M	M	G	M	M	M	G	M	P	General design does not indicate better emission characteristics than other piston engines.
Zinc-air battery, electric	G	G	G	M	G	P	M	P	M	M	G	P	Zero emissions except at power plant generating electricity for recharging. May be used in shopper-type vehicles.
Steam or vapor	M	M	M	M	M	G	M	P	P	G	M	M	Very poor from weight and space considerations. Low emissions, possibly in range of gas turbines.
Stirling cycle	M	M	M	M	G	G	M	M	P	G	M	P	Low emissions but poor space requirements and poor state of development.
Lead-acid battery, electric	G	M	G	M	G	P	M	P	P	P	G	M	Heavier, bulkier than zinc-air batteries.
Atomic, steam	P	P	G	M	M	P	G	P	P	G	M	P	Shielding problems add to the weight and space problems of the steam system.

G—Good; M—Medium; P—Poor

SOURCE: *Automotive News*, 27 June 1966.

TABLE 6.12
Energy and Power Characteristics

Secondary Batteries				
	Energy Density (w.-hr./lb.)		Power Density (w./lb.)	Remarks
Type	Theo-retical	Actual		
Commercial				
Lead—acid	70	8–13	32	
Lead—cobalt	—	18	23	
Nickel—iron	115	12–13	33	
Nickel—cadmium	99	15–20	40 *	
Silver—zinc	222	50	150	
Silver—cadmium	134	20–25	65	
Research and De-velopment				
Sodium—sulfur	345	150	100 150	Must be at 500° F.
Lithium—chlorine	1200	100	150	Must be at 1150° F.
Lithium—nickel fluoride	620	90–100, 150	20	Operates at room temperature
Zinc—air	464	80 50	20 50	Operates at 32° F. and above
Sodium—air	930	160–215	40–55	Operates at 265° F.
Gasoline—internal combustion engine	5850	600–1500	150–400	

* Bipolar type can reach 180 w./lb. in one-second bursts.

Steam or vapor cycles are heavy and bulky by today's standards, largely because of the boiler and condenser.[19]

Thus, there seems to be little threat to the supremacy of the gasoline engine for automobiles, although there are improvements to be made in the engines, and although there will be some minor chipping away of the gasoline market in selected applications by LNG, LPG, diesel and gas turbines, and later by electric vehicles. LPG and LNG offer advantages in the realm of lower hydrocarbon emissions, but the supply of LPG and LNG is not adequate for the entire automotive population and distribution would be hampered by the need for special storage equipment at each service station. With respect to NO_x emissions, these fuels offer no advantage over gasoline.

Truck and Bus

Consumption of truck fuel has expanded slightly faster than consumption of automotive gasoline. Trucks consumed 21.2 percent of all motor fuel in the United States in 1940 and in 1967 trucks consumed 27.6 percent of the motor fuel or 1.4 million barrels per day.[20] Buses represented an additional 1.2 percent of the total fuel consumption in 1967.

This expanding market uses primarily diesel engines although gas turbines are competing. Both types of power plants use the same fuel.

Figure 6.5 shows steady decrease in cetane number and sulfur content of average No. 1 and No. 2 Diesel Fuel in the United States since 1956. These trends are the result of increased cracking and hydrogenation. Increased hydrocracking (the newer, more severe hydrogenation procedure) will probably accelerate the trend to lower sulfur content and start a trend to higher cetane numbers.

Railroads

The changeover from coal- and oil-fired steam locomotives to diesel locomotives, and the steady decline in the use of electric rail propulsion, are shown in table 6.13. United States railroads use a quarter of a million barrels per day of diesel fuel.[20]

TABLE 6.13
Fuel and Power Consumed by United States Railroads, Class 1

	1950	1955	1960	1965	1968
Coal: 1,000 short tons	55,452	11,427	39	4	2
Fuel Oil: million gallons	2,284	376	89	77	42
Diesel Oil: million gallons	1,827	3,393	3,472	3,742	3,922
Electricity: million kw-hr.	2,260	2,082	1,641	1,509	1,317

SOURCE: United States Bureau of the Census, "Statistical Abstract of the United States: 1970," 91st ed. (Washington, D.C., 1970).

Because railroads use medium-speed diesels, the engines tolerate lower quality fuel than do buses and trucks. This is shown in figure 6.5.

A new stimulus has entered the railroad market with the current studies on high-speed commuter trains. The United Aircraft Turbo Train, developed by United Aircraft Corporation, applies a relatively new technology to railroading. Aircraft gas turbine engines drive the train through a helicopter-type transmission. Cars are constructed of lightweight aluminum alloys employing aircraft construction methods. Center of gravity is lower than for standard rail cars and the unique pendulum suspension system allows negotiation of curves at higher speeds. Two three-car train sets of this type are based in Providence, R.I., and run on the New Haven tracks between Boston and New York; five seven-car trains operate between Montreal and Toronto, Canada, on the Canadian National Railway tracks. Other gas turbine powered trains, such as the AiResearch-powered Budd Com-

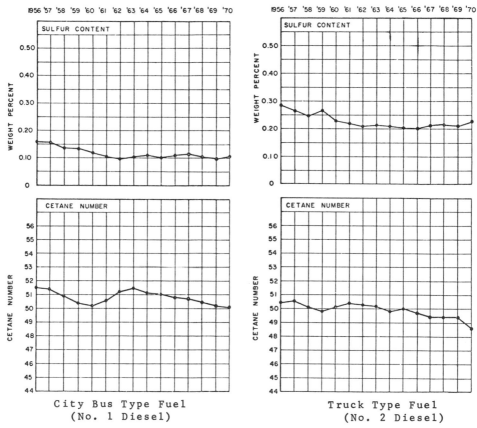

Fig. 6.5 Quality trends in diesel fuel in the United States.
SOURCE: E. M. Shelton and C. M. McKinney, "Diesel Fuel Oils, Winter 1969–1970," *Petroleum Products Survey No. 65,* Mineral Industries Surveys, United States Department of the Interior, Bureau of Mines, Washington, D.C.

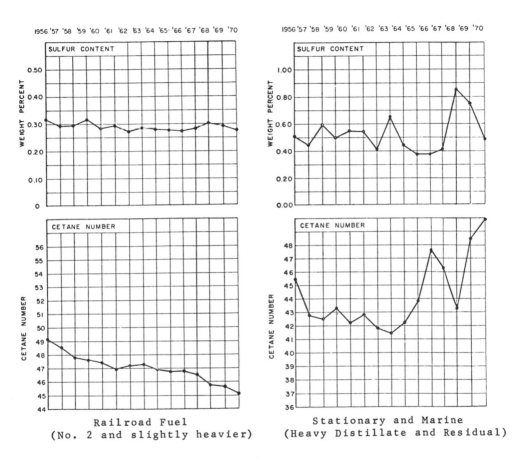

Railroad Fuel
(No. 2 and slightly heavier)

Stationary and Marine
(Heavy Distillate and Residual)

pany train, are also under development. Success with these projects could result in significant expansion of passenger train operation.[3]

Aircraft

Large numbers of gasoline engines for aircraft are manufactured, but they consume a declining fraction of the aviation fuel. Of the 23,000 aircraft engines produced in 1966, 92 percent were reciprocating. Conversely, only 12 percent of the 200 plus million barrels of commercial aircraft fuel consumed in North America in 1967 was aviation gasoline. In the larger sizes, the gas turbine has captured the aircraft propulsion market. This reflects the tremendous savings from reduced maintenance and improved reliability. American Airlines achieved an approved overhaul life of 11,000 hours on the PWA JT3D Turbo fan engine in May 1967. This is equivalent to three years elapsed time based upon an average aircraft utilization of 3,600 hours per year and is three times greater than the best overhaul life ever achieved with an aircraft piston engine.[3]

Figure 6.6 shows the growth in aircraft turbine fuel demand anticipated in 1980 in the United States.[3] This does not include Navy JP-5 aircraft jet fuel (65,000 bbl. per day in 1969) or the United States Air Force JP-4 fuel requirement (550,000 bbl. per day in 1969). JP-4 is about 40 percent naphtha or gasoline and 60 percent kerosene.

In a jet engine, fuel on the way to the burner is a coolant for hot lubricating oil, hydraulic oil, etc. Thermal stability at 400 F. is a common requirement and this level might be raised to 600 F. or higher for supersonic aircraft. For aircraft up to at least Mach 2.5, hydrocracked kerosene-type distillates can be made to meet this requirement at a modest price premium over present jet fuel.[21]

The United States Air Force Aero Propulsion Laboratory has projected the fuels shown in figure 6.7 for speeds up to Mach 14.[22] In the speed range of Mach 4 to Mach 10, it is anticipated that some fuels will absorb heat in endothermic decomposition reactions and the decomposition products will reach the burner nozzles in a gaseous state. The Air Force paper concludes as follows:

1. Present hydrocarbon fuels, such as JP-4, JP-5, and proposed JP-8 can be and are used at flight speeds up to about Mach 2.5.
2. Hydrocarbon jet fuels, such as JP-7, are currently available which are stable in the liquid state at temperatures approaching 700 F. and are capable of providing the necessary heat sink for flight speeds to about Mach 3.5. If additional precautions are taken in order to eliminate dissolved oxygen in the fuel, this type fuel could be used to flight Mach numbers up to about 4.5.

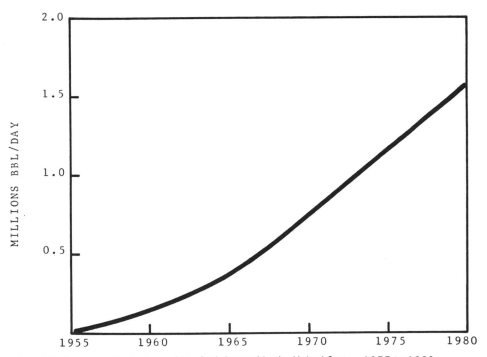

Fig. 6.6 Commercial aviation turbine fuel demand in the United States, 1955 to 1980.

3. Flight speeds beyond Mach 4 will require vaporizing-endothermic fuels for increased heat sink capacity. Fuels, such as methylcyclohexane, decalin and dicyclohexyl or mixtures thereof, appear to have the capability of providing sufficient cooling for propulsion systems up to speeds of Mach 9–10. At higher flight speeds liquid hydrogen fuel will be required.
4. For special applications requiring high volumetric heat content, high density hydrocarbons and boron slurry fuels have been developed and are being further evaluated in propulsion systems.

Marine

In 1939, 45 percent of the world shipping tonnage was coal-fired, but by 1966 this had dropped to 2 percent. Between 1956 and 1966, the world's fleets grew from 105 to 170 million tons; the percentage of motor vessels (primarily slow-speed diesels) moved up from 36 to 56 percent, while steam turbine-driven vessels decreased from 56 to 42 percent.[23]

Of the 1000 or more ships built each year, the majority over 2000 deadweight tons employ power plants of 5000 to 20,000 HP. Of the 2000 ton and larger ships completed in 1968, 94 percent used diesel propulsion, mostly slow-speed, direct drive machinery burning blends of residual fuel.[24] Single engines of 48,000 brake horsepower are available. *Motor Ship* magazine, January 1969, discussed the trend to ships of 60,000 to 100,000 HP, a range in which nuclear power will receive serious consideration.

Nuclear-powered submarine tankers have been proposed to carry 170,-000 ton cargoes of Alaskan oil.[25] Another proposal covers a 1400 ft. "Nuclear Eel" to carry 100, 50-ton railroad cars at a speed of 40 or 50 knots (with 500 or 1000 Mw of reactor power) on regular schedules between Europe and North America.[26] Edward Teller stated that marine power requirements are not in a truly economical range for reactors and that ships should stay with fossil-fuel plants except for special situations such as naval vessels.[27] However, some of these proposals promise to materialize as the best way to get the job done, even though they do not represent the best way to run a reactor.

Slow-speed marine diesels in the early 1950s burned distillates. Tests were soon started on Bunker C, which cost 40 percent less. Wear, corrosion, and maintenance problems multiplied and operators compromised on blends which give the optimum balance between fuel and maintenance costs. The fuel supplier must be prepared to mix in any proportions at his dock, but the most popular blends contain 8 or 10 percent distillate, which reduces the viscosity from 3500 Redwood No. 1 Seconds at 100° F. to 1000 or 1500. World consumption of bunker fuel oils was 400 to 450 million barrels in 1966 and increases 3 percent per year.[28] This includes not only blends but

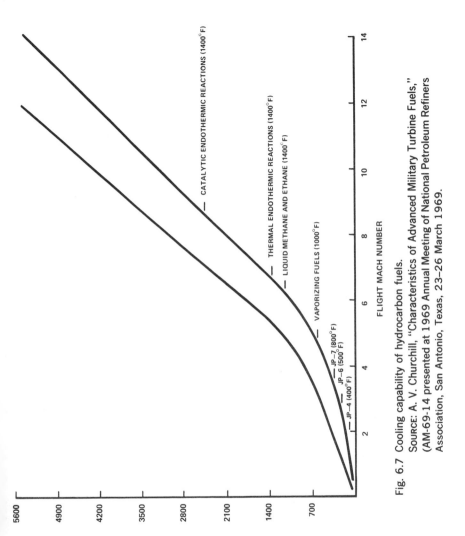

Fig. 6.7 Cooling capability of hydrocarbon fuels.

SOURCE: A. V. Churchill, "Characteristics of Advanced Military Turbine Fuels," (AM-69-14 presented at 1969 Annual Meeting of National Petroleum Refiners Association, San Antonio, Texas, 23–26 March 1969.

143

also distillate fuel for shipboard auxiliaries and for starting, stopping, and maneuvering operations of the main propulsion engines. Two recent developments presage expansion of the distillate market at the expense of Bunker C:

1. The United States Navy is adopting Navy Distillate, MIL-F-24397, in place of Navy Special (residual) as the main propulsion fuel for the fleet. This is a combination boiler, diesel, and gas turbine fuel, and by 1973, 50 million barrels per year will be required. Although the Navy's fuel bill will increase by 60 million dollars, a net savings of 130 million is anticipated from reduced maintenance and increased boiler efficiency. The Royal Canadian Navy adopted a similar plan some time earlier and now has several ships equipped and authorized to use "low-ash fuel," specification 3GP11.

2. Gas turbines are making a serious bid to power merchant as well as naval vessels. Distillate fuel is used in most current tests. Ship propulsion by industrial gas turbines burning Bunker C Fuel Oil was demonstrated successfully fifteen years ago, but operators showed little interest at the conclusion of these pioneering tests. General Electric is again offering industrial gas turbines for this service, and with the benefit of fifteen more years of development, they are supplying a power plant to burn selected residual fuel in a dry cargo ship for Broken Hills Proprietary, Ltd., Australia.

Gas turbines in marine service grew from 200,000 HP in 1960 to 3,340,000 by mid-1967; 90 percent for propulsion and 10 percent for auxiliaries. These are used on about 200 ships and craft, as compared with 13,785 military and 19,361 merchant vessels in the world (table 6.14). Installations have been primarily in gunboats, frigates, and patrol boats plus some hydrofoils and other high performance craft. Some diesel-powered military craft use gas turbine boost power to obtain exceptional speed in emergencies. The "Admiral William M. Callaghan," a 672 ft., 24,000 ton freighter powered by two 20,000 HP aircraft-type gas turbines, has been in trans-Atlantic service for two years with an appetite for 800 bbl. of distillate per day.

Most large container ships on order will have 32,000 to 36,000 HP, which dictated the use of steam when orders were placed.[29] Gas turbines are being specified for a few container ships because of their small size and weight and low requirement for supervision and on-board maintenance.

To summarize, steady growth in bunker fuel demand is expected because of increased tonnage and speed of vessels. Distillate demand will increase faster than the demand for residual, but diesels will power the bulk of world shipping for the foreseeable future. Part of this dominance arises from the continuing progress in design, including significant increases in power out-

TABLE 6.14
Marine Vessels, Worldwide

	United States	World Total
Military, 1969 *		
Aircraft carriers	56	76
Submarines	190	842
Communications, cruisers, command, flagships	44	95
Destroyers, frigates, escorts	616	1,422
Patrol, corvette, torpedo boats	63	2,295
Mine layers, sweepers, hunters	183	1,596
Landing ships and craft	276	1,047
Supply and auxiliary	586	1,617
Other and miscellaneous	1,346	4,795
Total	3,360	13,785
Merchant, 1000 tons and over, 1968 *		
Combination passenger and cargo	195	989
Freighters	1,511	11,868
Bulk carriers	53	2,609
Tankers	312	3,895
Total	2,071	19,361
Merchant, 2000 tons and over, completed in 1970 †		
Motor	1	1,043
Steam	11	102
Full container ships on order, 1 January 1971 ‡	—	127

* *Statistical Abstract of the United States, 1970.*
† *The Motor Ship,* January 1970, pp. 435–55.
‡ *"Motor Ship's* Survey of Ships on Order," *Motor and Steam,* December 1970.

put, reductions in specific weight, and improvements in reliability and ease of maintenance.

Pipelines

Pipelines, as "transporters of energy," moved impressive quantities in the United States alone in 1969.

Gas:	19.1 trillion cu. ft.	19.7 quadrillion Btu.
Crude:	5.0 billion bbl.	29.1 quadrillion Btu.
Products:	2.7 billion bbl.	15.1 quadrillion Btu.[30]

Pipeline movements of petroleum oils exceed 2 trillion barrel-miles per year in the United States, or about 21 percent of all freight transport.[20] Pipelines are popular because they transport energy cheaply, as shown in table 6.15.

Pipeline shipments increase 7 or 8 percent each year and continued growth is projected, with gas deliveries exceeding 22 trillion cu. ft. in 1974 and liquid petroleum shipments exceeding 12.5 billion barrels annually by 1980 (see fig. 6.8). Fuel for pumping, which probably amounts to .5 or 1.5 percent of the energy transported, will increase accordingly.

TABLE 6.15
Energy Transportation Costs
(cents/million Btu. [100 miles])

Oil pipeline	0.6
Gas pipeline (800–1000 psi)	1.5
Coal (unitrain)	3.1
Extra-high-voltage electric lines (500 KV)	10.0

SOURCE: H. R. Linden, "Power and Energy," *Institute of Gas Technology, Gas Scope*, no. 5.

Transportation of ammonia, coal slurries, and residual fuel by pipeline is in the advanced experimental or early commercial stages and is expected to expand considerably in the next decade.

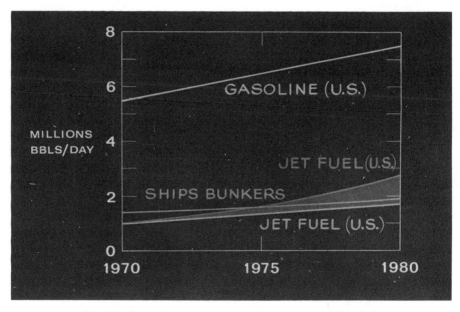

Fig. 6.8 Projections of growth trends for transportation fuels.

FUTURE FUEL REQUIREMENTS

Air pollution controls and new engines may alter the distribution of products from the petroleum barrel, but electric automobiles, nuclear ships, and other nonhydrocarbon systems are not expected to supply a large share of transportation in this decade. Gasoline will continue to be the dominant fuel for transportation because of its use in passenger cars. The power plant trends that will influence fuel requirements are summarized in table 6.16.

TABLE 6.16
Summary of Power Plant Trends Influencing Fuel Requirements

Future Automotive Engines	Nonautomotive Transportation		
Predictions	Predominant Now		Trend
1. Gasoline (modified) Still dominant (with pollution controls)	Ships	1. Diesel 2. Steam	1. Diesel 1. Gas turbine 3. Nuclear (few)
2. LPG (and LNG) for fleets City buses Taxis Delivery trucks	Truck and bus	1. Diesel	1. Diesel 2. Gas turbine
3. Gas turbines for prestige cars	Aircraft	1. Turbine 2. Gasoline	1. Turbine
4. Electric cars—shopper type	Railroads	1. Diesel	1. Diesel 2. Gas turbine (few)
	Pipelines	1. Natural gas engines 2. Diesels 3. Gas turbines	1. Natural gas engines 2. Diesels 3. Gas turbines

Reference notes

1. "On-Site Distillation of Diesel Fuel," *Diesel and Gas Turbine Progress* 36, no. 1 (January 1970): 48–49.
2. J. C. Winger, "The Petroleum Situation in November, 1969" (Energy Division, Chase Manhattan Bank, 5 January 1970).
3. R. C. Amero and R. P. Foster, "Impact of the Fast-growing Gas Turbine Market on Refinery Futures" (Preprint no. 16–69, 34th Midyear Meeting of the American Petroleum Institute's Division of Refining, Chicago, Ill., 12 May 1969).
4. D. H. Stormont, "US Hunger for Jet Fuel Hits 1 Million B/D Level," *Oil and Gas Journal* 67, 28 July 1969, p. 93.
5. R. P. Gilmartin, W. A. Horne, and B. R. Walsh, "Hydrogen Treatment Improves Fuel Oils," *Oil and Gas Journal* 54, no. 40 (6 February 1956): 85–88.
6. "North Slope Oil to Cut Prices, Relax Import Controls, Economist Tells NOFI Executives," *Fuel Oil News,* 34, no. 11 (November 1969): 32–33.
7. "Greater Use of Synthetic Fuels Appears Imminent," *Oil and Gas Journal* 39, no. 1 (January 1968): 23.
8. "Coal-to-Oil, Gas Processes: Closer but Not Commercial," *Chemical Engineering,* 3 November 1969.
9. P. A. Laurent, "Future Demand for Petroleum Products," *Chemical Engineering Progress* 65, no. 12 (December 1969): 29–38.
10. S. F. Jefferson and J. G. Trimmer, "The Changing Pattern of Product Demand," *Institute of Petroleum Review* 19, no. 224 (August 1965): 275–81.
11. Walter R. Hibbard, Jr., *Oil and Gas Journal* 65, no. 44 (10 October 1967): 33–44.
12. C. B. Kass, "Car Registrations and Metropolitan Traffic in 1980," Ethyl Corporation Report TS-145, June 1969.
13. "Gasoline to Stay Top Motor Fuel," *Oil and Gas Journal* 64, no. 47 (21 November 1966): 132.
14. A. H. Sweet, B. J. Steigerwald, and J. H. Ludwig, "The Need for a Pollution-Free Vehicle," *Journal of the Air Pollution Control Association* 18, no. 2 (February 1968): 111–13.

15. "Propane to Power a Police Fleet," *Butane-Propane News* 1, no. 4 (December 1969): 35–36.
16. F. V. Morriss, et al. "The Exhaust Content of Automobiles Burning Ethanol-Gasoline Mixtures," ACS Meeting Reprint no. 77 (September 1955).
17. E. S. Starkman, "Alternative Fuels for Control of Engine Emissions" (Paper 69-222, Air Pollution Control Association 62nd Annual Meeting, New York).
18. F. Jones, "Fuels for Fuel Cells," *Fuel* 48, no. 1 (January 1969): 47–60.
19. T. O. Wagner, "Practicality of Steam-powered Automobiles," (Paper no. 19-69, 34th Midyear Meeting of API Division of Refining, Chicago, Ill., 12 May 1969).
20. United States Bureau of the Census, "Statistical Abstract of the United States: 1970," 91st ed. (Washington, D.C., 1970).
21. R. P. Foster and E. M. Sutphin, "Economical Mach 3+ SST Fuel Available Now," SAE National Aeronautical and Space Engineering and Manufacturing Meeting (Los Angeles, Calif., 4–8 October 1965).
22. A. V. Churchill, "Characteristics of Advanced Military Turbine Fuels" (AM-69-14 presented at 1969 Annual Meeting of National Petroleum Refiners Association, San Antonio, Tex. 23–26 March 1969).
23. "Fuel for the World's Ships," *Petroleum Times* 71, no. 1821 (26 May 1967): 754.
24. R. P. Giblon, "A Few Points Relating to the Use of Gas Turbines by the Merchant Marine" (George C. Sharp, Inc., Report to Annual Gas Turbine Conference and Products Show, Houston, Tex., March 1967).
25. M. Barends, "Under the Northwest Passage," *Mechanical Engineering* 92, no. 2 (February 1970): 79.
26. R. L. Whitelaw, "The Nuclear Eel," *Mechanical Engineering* 91, no. 11 (November 1969): 23–27.
27. E. Teller, "Energy from Oil and from the Nucleus," *Journal of Petroleum Technology* 17, no. 5 (May 1965): 505–8.
28. "World Energy Supplies" (Estimated from Statistical Papers Series J, no. 11, Department of Economic and Social Affairs, Statistical Office of the United Nations, New York, 1968).
29. "Full-Container Ships on Order—January 1, 1969," *The Motor Ship* 49, no. 582 (1969): 469.
30. J. P. O'Donnell, "Pipeline Progress Report: Interstate Lines Set New Record," *Oil and Gas Journal* (8 June 1970): 115.

7. Air Quality Criteria

O. W. Stopinski and R. P. V. Boksleitner
Environmental Protection Agency

This presentation will reflect a federal orientation toward air quality and its achievement; we will consider air quality criteria as a basis for control, federal control alternatives, and current federal air quality proposals. There must be a concerted effort, however, on the part of the private as well as the public sector and among all levels of government if air pollution is to be controlled.

Since the terms air quality criteria and air quality standards will be used liberally in this presentation, definitions for our purposes will be useful. *Air quality criteria* are an expression of the scientific knowledge of the relationship between various concentrations of air pollutants and their adverse effects on man and his environment. Air quality criteria are descriptive; that is, they describe the effects that have been observed when the ambient air concentration of a pollutant has reached or exceeded a specific figure for a specific time period. In developing criteria, many factors have to be considered: the chemical and physical characteristics of the pollutants, the techniques available for measuring these characteristics, along with exposure time, relative humidity, and other conditions of the environment. One must also consider the contribution of all such variables to the effects of air pollution on human health, agriculture, materials, visibility, and climate. Further, the individual characteristics of the receptor must be taken into account.

Air quality standards are prescriptive. They prescribe pollutant exposures that a political jurisdiction determines should not be exceeded in a specified geographic area, over given time periods, and are used as one of several factors in designing legally enforceable pollutant emission standards.

A logical beginning to air resource management or air pollution control is the determination of the quantity of various pollutants which is not to be exceeded. One might consider two general bases for achieving a certain air quality: *technical feasibility* and *air quality criteria.* Technical feasibility accepts the premise that sources of air pollution emissions should be controlled to the highest degree reasonably feasible through the use of available technological means. In this case, technological capability defines how much pollution may be emitted into the air and, thus, the resulting air quality. A drive for the development of improved technology is rarely self-inspired; technological improvement generally requires external stimulation.

AIR QUALITY CRITERIA INDICATE AIR QUALITY STANDARDS

At the same time, air quality criteria are statements of dose and response of a given pollutant at various concentrations acting for various time periods on various receptors. In their most refined form, criteria could provide a checklist for those at any governmental level faced with the responsibility of establishing air quality standards. Ideally, one could scan the list of effects and identify those of concern in the jurisdiction for which standards are being promulgated. For instance, in a state such as Arizona noted as a haven for asthmatics, criteria on respiratory effects would be extremely pertinent while effects on Northern White Pine might deserve little attention. The identification of pertinent criteria, then, leads to the development of pertinent standards. Air quality standards which have been developed from air quality criteria can provide an objective basis for developing an air pollution control implementation program.

Under the provisions of the Air Quality Act of 1967 and now under the Clean Air Act as amended in 1970, the federal air pollution control program has prepared and issued air quality criteria documents for various air pollutants.

These documents are compilations of the latest available scientific information on the sources, prevalence, and manifestations of recognized air pollutants. Their most important function is to describe effects associated with, or expected from, air pollutant concentrations in excess of specific concentrations for specific time periods. These effects generally include visibility reduction, material and vegetation damage, economic costs, nuisances, and adverse effects on the health and well-being of man and animals.

Air quality criteria documents are planned to provide comprehensive information in the following major areas whenever pertinent: an environmental appraisal which includes origin (natural and man-made) and fate;

physical and chemical properties; spatial and temporal distribution in the atmosphere; and atmospheric alterations, which include chemical transformations and meteorological influences on the pollutant. Measurement technology is included so that a reference basis for quantitating the pollutant will be available.

The criteria present data on the effects on human health including laboratory animal data where human inferences exist. Toxicologic criteria, which include behavioral, sensory, biochemical, and physiological mechanisms and responses, and epidemiological and clinical criteria, which include field and clinical studies, are included. Criteria on biological and physical effects on natural and cultivated plant life, domestic and wild animal life, and materials are included. Criteria of social and aesthetic effects are included when information exists on these parameters. Economic impact, including that resulting from all other effects listed above plus effects on specific economic parameters such as real property values, are also included in the criteria document. The search for air quality criteria usually highlights gaps in knowledge. Therefore, a section on research needs is included.

LIMITATIONS TO AIR QUALITY CRITERIA

There are shortcomings in some of the data which are the basis for criteria:

1. Effects relating exposures to ambient air do not take into account the presence of all of the pollutants causing the net result; some of the pollutants may be intermediate or unstable products of known substances, while the presence of others may be unsuspected, thus not even monitored.
2. Laboratory experiments using simulated polluted atmospheres often cannot replicate the actual ambient air in composition, temperature, and humidity simultaneously. The presence of other pollutants (some not routinely measured) may also contribute to the observed effects.
3. It is extremely difficult, if at all possible, to state minimum or threshold levels for a particular pollutant with reference to a particular effect. Long-term exposures allow many variables to exert an additional influence on the outcome; short-term exposures often do not yield measurable effects when done at realistic ambient concentrations.
4. Many studies of effects are not directly comparable with each other because of nonparallel exposure times or conditions, and because of variations in measurement techniques and averaging times.
5. Air quality data are gathered at selected monitoring sites and may not

reflect the actual exposures of the subjects being studied. In addition, with certain pollutants, the ambient air concentrations are not the concentrations that actually reach the sensitive tissues.

Even though the accuracy and precision of data concerning health and welfare effects often leave something to be desired, until better measurements are possible, our actions must be based upon the best knowledge we have and be guided by the principle of enhancing the quality of the air environment and, thus, human life.

CLEAN AIR ACT

Federal control alternatives provided for in the Clean Air Act, PL 91-604, which require consideration of air quality criteria or equivalent technical reports, include:

1. Section 108, which provides for the issuance of air quality criteria and control techniques documents
2. Section 109, which provides for the proposal and promulgation of National Primary and Secondary Ambient Air Quality Standards for those pollutants for which criteria and control technique documents have been issued
3. Section 110, which provides for the implementation of actions designed to attain the national standards
4. Section 111, which provides for the proposal and promulgation of standards of performance
5. Section 112, which provides for the designation of hazardous air pollutants and the proposal and promulgation of emission standards for these
6. Section 202, which provides for the promulgation of emission standards from mobile sources
7. Section 211, which provides for the control of fuels and fuel additives
8. Section 231, which provides for the study of the effect of aircraft emissions on air quality and the ultimate promulgation of aircraft emission standards
9. Section 601 of the Federal Aviation Act, which was amended by the Clean Air Amendments of 1970

Of these provisions, five provide new avenues for control of air pollutants: (1) Section 111, standards of performance; (2) Section 112, hazardous air pollutants; (3) Section 211, fuel additives; (4) Section 231, aircraft emissions; and (5) Section 601, FAA control of composition of aircraft fuels. Sections 108 and 109 go hand in hand. Section 108 provides for the development and issuance of air quality criteria while section 109 requires the administrator of the Environmental Protection Agency (EPA) to include with the issuance of air quality criteria proposed national primary and secondary air quality standards.

Interested parties may comment on the proposed standards. Standards are to be promulgated ninety days after the proposed standards are published.

Air quality criteria documents, as initially issued under the provisions of the Clean Air Act, as amended in 1967, were published to serve as guides to the states in setting ambient air quality standards. The state adopted standards were then reviewed by the Federal Air Pollution Control Office.

Today air quality criteria serve the same function—a basis for the establishment of ambient air quality standards; the only difference is that now the administrator of the EPA establishes the standards, which must apply throughout the country. At the same time the states may develop their own air quality standards. Such state adopted standards must be at least as stringent as the national standards.

Ambient air quality standards as defined in the Clean Air Act are of two types:

1. Primary Ambient Air Quality Standards are those ". . . which in the judgment of the Administrator, based on such criteria and allowing an adequate margin of safety, are requisite to protect the public health."
2. Secondary Ambient Air Quality Standards ". . . shall specify a level of air quality the attainment and maintenance of which in the judgment of the Administrator, based on criteria, is requisite to protect the public welfare from any known or anticipated adverse effects associated with the presence of such air pollutant in the ambient air."

Once federal standards are promulgated, Section 110 of the Act requires each state to submit plans to implement the standards within nine months after the promulgation of National Air Quality Standards. Primary standards are to be met within three years of the approval date of the plan. This section does provide for extension of the time limit and for federal implementation where a state does not submit a suitable plan or does not act to implement the standards. This section also provides for judicial review in a United States court of appeals.

Section 111 provides for the establishment of emission standards for new sources by the federal government and for existing sources by state governments. These emission regulations would apply to categories of sources designated by the administrator of the Environmental Protection Agency. Enforcement may be delegated to states.

Section 202 is part of the National Emission Standards Act which, beginning in 1965, gave the federal government authority to regulate motor vehicle emissions. The control of emissions from *new* motor vehicles is preempted by the federal government, except in states having adopted motor vehicle standards prior to 30 March 1966—essentially, this applies to California.

Section 211 provides for the regulation of fuels. The administrator may, by regulation, designate any fuel or fuel additive and, after such date prescribed, no manufacturer or processor of any such fuel or additive may sell, offer for sale, or introduce into commerce such fuel or additive unless the administrator has registered it. The administrator may prohibit or otherwise control the sale or use of fuels found to endanger public health or welfare or which significantly interfere with emission control devices.

Section 231 provides for the study of the effect of aircraft emissions on air quality and the ultimate promulgation of aircraft emission standards.

Section 601 of the Federal Aviation Act, as amended, provides for control of the composition of aviation fuels by the administrator of the FAA for the purpose of controlling or eliminating aircraft emissions which the administrator of the EPA may determine endangers the public health.

All of these regulations generally provide for notice of proposed regulation in the *Federal Register* with subsequent time for submission and consideration of comments from interested parties. In some cases, public hearings are required.

ADMINISTRATIVE ACTIONS

Following the legal enactment of the Clean Air Amendments on 31 December 1970, several actions have been taken by the administrator of the EPA in accordance with various provisions of the Act. On 30 January 1971, a list of pollutants for which the administrator planned to issue air quality criteria was published. The list consisted of a single pollutant category—nitrogen oxides. At the same time, notice of issuance of the document, Air Quality Criteria for Nitrogen Oxides, was announced.

Simultaneously announcement was made by the administrator of proposed National Primary and Secondary Ambient Air Quality Standards for six pollutants. Criteria documents for five of these pollutants had been previously issued under provisions of the Clean Air Act, as amended in 1967. Accordingly standards were proposed for sulfur oxides, particulate matter, carbon monoxide, photochemical oxidants, hydrocarbons, and nitrogen dioxide.

The National Primary and Secondary Ambient Air Quality Standards, proposed 31 January 1971 and adopted 30 April 1971, are:

1. Sulfur oxides as sulfur dioxide
   Primary standards
   a) 80 $\mu g/m^3$ annual arithmetic mean
   b) 365 $\mu g/m^3$ maximum 24-hour concentration not to be exceeded more than once per year

   Secondary standards
   a) 60 $\mu g/m^3$ annual arithmetic mean
   b) 260 $\mu g/m^3$ maximum 24-hour concentration not to be exceeded more than once per year
   c) 1300 $\mu g/m^3$ maximum 3-hour concentration not to be exceeded more than once per year

2. Particulate matter
   Primary standards
   a) 75 $\mu g/m^3$ annual geometric mean
   b) 260 $\mu g/m^3$ maximum 24-hour concentration not to be exceeded more than once per year

   Secondary standards
   a) 60 $\mu g/m^3$ annual geometric mean
   b) 150 $\mu g/m^3$ maximum 24-hour concentration not to be exceeded more than once per year

3. Carbon monoxide
   Primary and secondary standards
   a) 10 mg/m^3 maximum 8-hour concentration not to be exceeded more than once per year
   b) 15 mg/m^3 maximum 1-hour concentration not to be exceeded more than once per year

4. Photochemical oxidants
   Primary and secondary standards
   125 $\mu g/m^3$ maximum 1-hour concentration not to be exceeded more than once per year

5. Hydrocarbons †
   Primary and secondary standards
   125 $\mu g/m^3$ maximum 3-hour concentration (6 to 9 A.M.) not to be exceeded more than once per year

6. Nitrogen dioxide
   Primary and secondary standards
   a) 100 $\mu g/m^3$ annual arithmetic mean
   b) 250 $\mu g/m^3$ 24-hour concentration not to be exceeded more than once per year

On 31 March 1971, a list of hazardous air pollutants was published in the *Federal Register*. This list was comprised of three materials: asbestos, beryllium, and mercury. Simultaneously, a list of source categories for which the administrator plans to propose emission standards within the ensuing 180 days was published. The five source categories are: steam-electric power plants, municipal incinerators, cement plants, sulfuric acid plants, and nitric acid plants.

On 31 March 1971, in accordance with requirements of Section 231 of the Act, the Air Pollution Control Office of the EPA initiated a study of aircraft emissions, with proposed emission standards to be published in the *Federal Register* within 180 days thereafter.

In addition to the proposed aircraft emission standards aforementioned, other actions are pending in the relatively near future. The administrator plans to issue guidelines for the formulation of implementation plans designed to attain the air quality specified by the air quality standards. These implementation plans, to be adopted by the states, must be submitted to the administrator within nine months following the promulgation of the national standards.

The Air Pollution Control Office of the Environmental Protection Agency is endeavoring to do its part to insure air quality that will not be detrimental to health or welfare. It is our hope that all people involved in the use of energy—and that is all of us—will be mindful of the pollution produced by energy use or conversion and will consider it an important responsibility to minimize all forms of pollution.

8. Water Quality Criteria

Harold W. Wolf
Professor, Texas A. & M. University

Our present life-style and economic growth cannot continue without expending vast amounts of energy. In the decade of the sixties, when the population of the United States grew by almost 10 percent, the total energy consumption nearly doubled. Extrapolating current trends, the Battelle Memorial Institute forecasts a usage of 170,000 trillion Btu. by the year 2000 as compared to 62,000 trillion Btu. utilized in 1968. This forecast, however, does not reflect the growing concern for protecting the environment and the resultant changes it may create in energy production and consumption.

If increasingly stringent protective environmental measures must be adopted, then what parameters or criteria shall be employed in decision making? Two of the more important areas of concern in the generation of energy are radioactive wastes and heat rejection.

THERMAL CRITERIA

A major impetus to setting water quality standards was the Federal Water Quality Act of 1965. That act encouraged states to establish water quality standards for interstate streams and coastal waters by 30 June 1967. The standards for all states and other jurisdictions have been approved by the Secretary of the Interior, although the approvals have been with reservations in some cases. Many of the reservations relate to temperature criteria.

The heat rate, or the amount of heat required by a plant to produce one kilowatt-hour of electricity, is a function of the temperature, pressure, and quality of the steam; for fossil-fuel plants, the heat rate depends upon the metallurgical properties of construction materials to retain their strength in fireboxes and boilers when exposed to very high temperatures and corrosive hot gases. As the temperature, pressure, and quality of the steam increase, more heat energy is carried to the turbine by each pound of steam and, therefore, the plant's efficiency increases. The efficiency of nuclear plants, of the boiling-water reactor type, being constructed or planned to 1975 will be less than the fossil-fuel plants of the early 1970s. More efficient nuclear plants are expected, but they are unlikely to come into operation until the mid- or late-1980s.[1]

In all steam-electric power plants operating or in the design stage in the early 1970s most of the waste heat is carried from the plant by cooling water. Either the cooling water is used on a once-through basis and returned to the body of water which will be enhanced in temperature or it passes through various types of air-water contact cooling devices.

A typical 40 percent efficient fossil-fuel power plant that requires 9,000 Btu. per kilowatt-hour of electricity produced discharges 4,237 Btu. of waste heat to the environment. Currently, a nuclear power plant operating at 33 percent efficiency with a heat rate of 10,342 Btu. per kilowatt-hour would discharge 6,400 Btu. of waste heat, or nearly 50 percent more heat than a fossil-fuel plant.[2]

A recent report of the Library of Congress for the use of the Joint Economic Committee of the House and Senate notes that waste heat discharged to waterways does not directly affect the public health, but it can markedly change the quality of the water for further use and also can drastically affect aquatic life.[2] Some individuals see these effects as being only undesirable; others see beneficial aspects.

Most living organisms require oxygen to maintain their life and reproductive functions. Metabolic activity is temperature-dependent and can double or triple over a 10° C. rise. The amount of dissolved oxygen in water decreases as the temperature increases. Therefore, the supply of oxygen decreases just as the biochemical oxygen demand rate is increasing.

Within limits, increased water temperatures can cause more rapid development of fish eggs, spat, fingerlings, and even some larger fish. Beyond specific temperatures, however, hatch is reduced and mortality in the development stages is higher.

A particular problem associated with increased water temperatures for migratory fish has been noted. Changes in metabolic rates caused by temperature rises have started premature spawning or migration.

Increased water temperature results in accelerated evaporation rates, thereby concentrating mineral salts in the water. In northern areas, heated water can reduce ice cover, at least locally, resulting in an improved oxygen budget by keeping the surface open to absorb oxygen from the air. Deleterious effects include ground level fog and drizzle on the water and adjacent land areas.

Increased temperature can make cold waters more desirable for water contact sports or warm waters less desirable. Similarly, heated discharges can reduce the value of water for industrial cooling purposes.

The preceding observations have been taken from the Library of Congress report. The Department of the Interior National Technical Advisory Committee oh Water Quality Criteria states the following concerning heat rejection:

> . . . In some localities and at certain times, elevation of temperature may be desirable . . . , but in most cases total recreational values . . . are more likely to be reduced than enlarged. . . .

> Excessively high temperatures may lessen the pleasure of some water contact sports. High temperatures limit the dissipation of body heat and may, through elevation of the deep body temperature, produce serious physiological disturbances. It has been determined that a person swimming expends energy at the rate of approximately 500 calories per hour. This is about five times the rate when sitting still and about twice the rate when walking. This energy must be dissipated to the environment to avoid a rise in the deep body temperature. When conduction is the principle means of heat transfer from the body and exposure to the environmental conditions is prolonged, 90° F is the approximate limit for persons expending minimal energy. Since most swimmers utilize energy at a moderate rate, the maximum water temperature that will not induce undesirable physiological effects after prolonged exposure must be less than 90° F. Experience with military personnel exposed to warm water continuously for several hours indicates that 85° F is a safe maximum limit.

> Limited exposure to water warmer than 85° F can be tolerated for short periods of time without causing undesirable physiological effects. In fact, some people get particular enjoyment from bathing in water from hot springs. However, these are special circumstances and persons bathing in such water usually limit their exposure to short periods of time and do not engage in moderate exercise such as swimming for prolonged periods.[3]

It is evident from this discussion that interpretation of temperature effects differs depending on the source. Using these criteria the states have produced individual temperature standards in response to the Water Quality Act. In reviewing temperature standards for waters used for fish propagation, Louisiana was found to have the highest value, 98.6° F. Some states stipulate several temperatures. Nebraska, for instance, specifies 65° F. for trout streams and 90° F. for warm-water fisheries. Maryland has a temperature limit of 90° F. for tidal waters and 93° F. for nontidal waters. The 90° F. limit in Maryland was adopted to apply to the discharge of the Calvert Cliffs Atomic Power Plant; this brings us to our second major concern: radioactivity in water.

RADIOACTIVITY CRITERIA

All criteria for radiation protection are derived from guidelines developed by the National Council on Radiation Protection and Measurement (NCRP). The NCRP was organized in 1929 in an effort to provide the United States with a unified voice at meetings of an already existing international group, the International Commission on Radiological Protection (ICRP).

At the first meeting of the ICRP in 1928, competing delegations from two American radiological societies had each claimed to be the authoritative body representing this country. As a result, no agreed upon American position was possible and recommendations prepared by a British protection committee were adopted by default.

Over the next several decades, the NCRP and ICRP held an almost undisputed position in the setting of radiation standards used to protect workers and the general public in the United States. The NCRP's recommendations were issued in handbooks bearing the name of the National Bureau of Standards (NBS) in accordance with its (the NBS's) statutory responsibility to cooperate with other governmental agencies and with private organizations in the development of standard practices, incorporated in codes and specifications.[4] It was not until 1959 that the government suddenly realized that it was relying primarily on private organizations to determine acceptable radiation protection standards. Hence, a new governmental organization, the Federal Radiation Council (FRC), was established to promulgate more "official" guidelines. The FRC—which came to consist of the heads of AEC, Defense, Commerce, Labor, Health, Interior, and Agriculture—was empowered to recommend standards which, after promulgation by the president would become official guidance for all federal agencies. The FRC was recently abolished and its functions were transferred to the new Environmental Protection Agency (EPA), and now the government will promulgate standards of its own through the new EPA.[5]

Existing standards must not be dismissed with impunity as representing the self-interests of proprietary industries or groups. Indeed, they derive from an impressive array of prestigious national and international organizations. A most controversial standard today is one which stipulates that the radiation dose received by the general population should not exceed an average of 170 millirems per person per year exclusive of medical exposures and natural background radiation. This is the standard which has been used in calculations of the number of deaths that would allegedly result if the general public actually received this permissible dose.[5]

The literature of virtually all the standards groups is laced with warnings that the standards involve value judgments and that the final decisions should be made by society, but thus far society has not come to grips with this complex problem and scientists have been left in charge by default.[5] Barry Commoner has stated that the determination of an acceptable balance between benefits and risks is a value judgment that should involve the public. Harold P. Green, an attorney, urges greater public participation in the risk/benefit determination maintaining that risk/benefit decisions are political concerns and not purely scientific problems. They should be debated in the rough and tumble of the political process. "What benefits does the public want and what risks is it willing to assume?" [6] Since the various standards groups have refused to get involved in estimating how many deaths might result if the public received the radiation allowed by the standards or in trying to quantify the presumed benefits of atomic energy, the public is left with little more than an assurance that the risk is acceptable.[5]

It is not surprising, therefore, that the standards are undergoing intensive governmental review by the new EPA, their first review in more than a decade. And without a doubt there will be some changes made, one of which may be an additional limit in Drinking Water Standards for tritium.

In reviewing various state responses to the Water Quality Act, one finds that standards for radioactivity vary also. Some cite *Drinking Water Standards* (Idaho), others cite the *NBS Handbook* 69 (New Mexico), and some utilize prose, such as "Concentrations shall not be harmful to aquatic life or result in accumulations in edible plants and animals that present a hazard to consumers" (Guam).

One of the publicized critics of the AEC, John Gofman of the Lawrence Radiation Laboratory, testified recently on the Calvert Cliffs nuclear power development.[7] He is reported as saying that the AEC, in arriving at standards, fails to take adequate account of the biological concentration of radiation in animals and, I presume, in plant life as well. There is, indeed, a mass of scientific literature that deals with this concentration aspect.[8] The highest concentration figure observed in *Water Quality Criteria* the California publication which is now somewhat out of date, shows a factor of 2,040,000 for Cerium 141 and 144 in algae.[1] Certainly concentration is a factor to be concerned about and should be taken into consideration. But perhaps this phenomenon could be put to productive use in protecting the environment.

ENVIRONMENTAL COST OF POWER PRODUCTION

Most of the world's production and consumption of energy during its entire history has occurred during the last twenty years.[6] Providing fuel to generate the quantities of energy needed in just the next thirty years will pose a substantial problem for the energy industries and for government policy, since the nation has been consuming its higher grade, more accessible resources first.[2] Projections of past and future production of fossil fuels show that the maximum length of time that fossil fuels can serve as a major energy resource can barely exceed three centuries.[6] As Paul Ehrlich remarked, when fossil fuels near depletion, what will power the tractors and other machines man requires for his feeding?

I am reminded of a speech made in 1898 by the distinguished Sir William Crookes who said, "The bread supply of the world [is] threatened with exhaustion . . . unless something develops to take the place of the rapidly diminishing supply of nitrogenous fertilizers." [8]

The report of the Library of Congress for the use of the Joint Economic Committee which I have drawn on so extensively notes that

1. Even the much smaller energy consumption of the last three decades has already created serious environmental problems.
2. The recognition of finite abilities for the air and water and landscape to yield fuels and assimilate wastes poses a direct challenge to growth.
3. There is some doubt that objectionable environmental effects of large steam-electric plants can be corrected.
4. The recent record of offshore development of petroleum reserves suggests a "go slower" policy.
5. The ratio of proved reserves to production of natural gas declined for the first time in 1968 and again in 1969.[2]

These observations would appear to indicate that we as a nation cannot continue to exploit our resources as we have in the past. We must expand our systems concepts to include the cost of depleted resources as well as reuse of the used products. Dr. DuBridge, President Nixon's science advisor, stated it this way:

It may be that energy consumption is growing so fast in part because the price does not include the full cost to society of producing and delivering it. I believe that efficient power production is just as important as ever to our economic growth, but we delude ourselves and perhaps short-change future generations when the price of electricity does not include the cost of the damaging impact its production imposes on the air, water, and land. If the total social cost of electricity or other products are included in its price, consumers will have the inherent ability to consider the effect of their decisions on the environment.[9]

Reference notes

1. J. E. McKee and H. W. Wolf, *Water Quality Criteria* 2d ed., publ. 3A (Sacramento, Calif.: State Water Quality Control Board, 1963), p. 343.
2. *The Economy, Energy, and The Environment,* Environmental Policy Division, Legislative Reference Service, Library of Congress (United States Government Printing Office, 1 September 1970).

3. *Water Quality Criteria* (Report of the National Technical Advisory Committee to the Secretary of the Interior, Federal Water Pollution Control Administration, United States Government Printing Office, 1 April 1968).
4. National Committee on Radiation Protection, *Maximum Permissible Body Burdens and Maximum Permissible Concentrations of Radionuclides in Air and in Water for Occupational Exposure*, National Bureau of Standards Handbook 69 (5 June 1959).
5. P. M. Boffey, "Radiation Standards: Are the Right People Making Decisions," *Science* 171 (26 February 1971): 790–93.
6. L. Loevinger, "States' Rights in Radiation Control," *Science* 171 (26 February 1971): 790–93.
7. K. Scharfenberg, "Calvert Plant Radiation Called Unsafe," *The Washington Post,* 8 December 1970.
8. Sir William Crookes (In a presentation of the "Wheat Problem," at which time the author was elected President of the British Association for the Advancement of Science, September 1898).
9. Testimony before the Subcommittee on Intergovernmental Relations, Senate Committee on Government Operations, 3 February 1970.

9. Present Fossil Fuel Systems and Their Emissions

A. L. Plumley
Combustion Engineering, Inc.

> *Be it known to all within the sound of my voice whosoever shall be found guilty of burning coal, shall suffer the loss of his head.*
>
> *. . . look you, this brave o'erhanging firmament, this majestical roof fretted with golden fire, why, it appears no other thing to me but a foul and pestilent congregation of vapors.*
>
> *We have met the enemy and he is us.*

Each of the above quotations is a vocalization of the frustration of man upon the realization that his environment is becoming overloaded with waste. It appears from the first quotation that King Edward I of England, in 1276, dealt with offenders in a fashion that we would consider unacceptable today. The second, Hamlet's description of Elsinore castle written by Shakespeare in 1600, is applicable to urban America of the twentieth century. Indeed, the third statement, made by the comic strip character Pogo and recently quoted by the keynoter of an air pollution control conference, is a succinct recognition of the fact that technological man has provided the means of polluting the atmosphere—but he also holds the key to minimizing emissions.

Air pollution is not a new problem, but the need for both immediate and long-range solutions has accelerated rapidly. The recently passed National Air Quality Standards Act of 1970, together with growing public awareness and the increased willingness of industry to pioneer emission control measures, promises a dramatic improvement.

Emissions from combustion of fossil fuels may be generally classified as from transportation or stationary sources. In table 9.1, the relative contribution from various sources is indicated.[1] Combustion of fossil fuels in stationary sources accounts for an annual emission of about 9 million tons of particulate matter, over 24 million tons of sulfur oxide, and 10 million tons of nitrogen oxide.

TABLE 9.1
Estimated Nationwide Emissions, 1968
(10^6 tons/year)

Source	CO	Particulate	SO_2	HC	NO_x
Transportation	63.8	1.2	0.8	16.6	8.1
Fuel combustion in stationary sources	1.9	8.9	24.4	0.7	10.0
Industrial processes	9.7	7.5	7.3	4.6	0.2
Solid waste disposal	7.8	1.1	0.1	1.6	0.6
Miscellaneous	16.9	9.6	0.6	8.5	1.7
Total	100.1	28.3	33.2	32.0	20.6

SOURCE: *Nationwide Inventory of Air Pollution Emissions—1968,* National Air Pollution Control Administration Publication AP-73 (August 1970).

TYPES OF EMISSIONS

There are three classes of emissions from power boilers that are significant from an air pollution standpoint: (1) particulate matter, (2) sulfur oxides, and (3) nitrogen oxide. Historically, the emission of particulate matter has received the greatest attention with regard to air pollution control since it is readily observed and is a source of public complaint on both a nuisance and an aesthetic basis. Present technology is adequate to significantly reduce particulate emission; however, improvements in both technology and economics of control can be expected.

Sulfur dioxide is currently receiving major attention because of possible adverse health effects and demonstrated damage to vegetation. Oxides of nitrogen are also receiving considerable attention because they participate in the complex series of chemical reactions in the atmosphere that lead to the formation of photochemical smog.

Particulate Emission

An estimated 28.3 million tons of particulate matter was emitted in the United States in 1968.[1] This figure is expected to increase to 30 million tons for 1970.[2] Figure 9.1 shows that about 33 percent, or 9 million tons, is attributable to the combustion of fossil fuels. Of this amount, 5.6 million

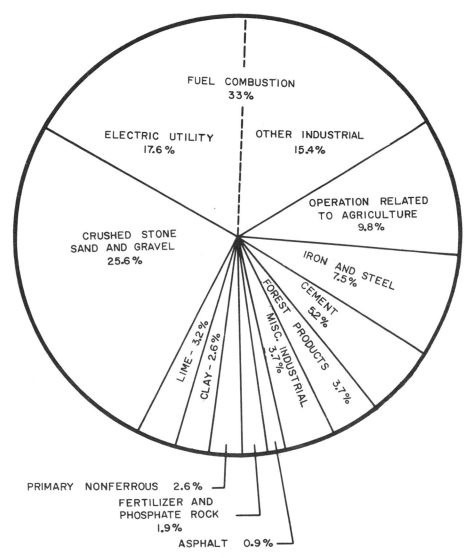

Fig. 9.1 Nationwide sources of particulate emissions, 1970.

tons was emitted from power plants and 2.6 million tons by industrial sources. The remainder was contributed by mobile sources.

Boiler particulate emissions have been gradually reduced over the years by improvements in the combustion process. However, the control has been more difficult because the quantity of particulate is less and, therefore, more difficult to remove. Regardless of the method of firing, there are variations in the particulate emissions due to the type of fuel fired.

Gas-fired boilers are relatively free of particulate emissions. If the effluent gas includes particulates, it is usually a sign of improper combustion.

Oil-burning units do emit particulate matter. The emission loading is very light (about 0.03 grains per standard cubic foot) and would be acceptable if the emissions were not made up of very fine particles. These fine particles cause the stack discharge to be quite visible because the particle size approaches the wave length of visible light. Although the quantity of particulate matter emitted is well within the dust loading requirements of air pollution ordinances, the visibility of the plume can be a source of complaints.

At present, studies are not complete and there are conflicting reports regarding particle size range when firing oil; the reports vary from about 0.1 microns to over 10 microns. In samples collected from a series of operating units, size distribution studies showed an average of 96 percent of the particles emitted from large oil-burning units had diameters less than 3 microns and 80 to 95 percent were less than 1 micron.[3]

Emissions from coal-fired boilers vary considerably depending on the ash content of the coal and the type of firing. For a coal with 10 percent ash, most systems have a dust loading in the range of 1 to 5 grains per standard cubic foot.

Particulate size distribution is extremely important. It can dictate the use of certain collection equipment and affect its design. To help understand the variation in particle size with the method of firing figure 9.2 is presented.[4] In this figure, the left-hand curves represent fly-ash size distribution for stokers. As might be expected, pulverized coal fly ash is further to the right, with about 45 percent by weight of the dust below 10 microns. On the extreme right, the size distribution of dust from a cyclone furnace is presented. This figure is, at best, an approximation. It is very difficult to arrive at average size distributions for different methods of firing coal, since much depends on the fuel preparation, excess air used, firing rate, and furnace design, as well as the method of firing.

It should be noted that nearly all fly ash under discussion is smaller than 44 microns. A further understanding of the size problem may be seen in figure 9.3. In the size range of concern, the settling rate is less than one-half foot per second.

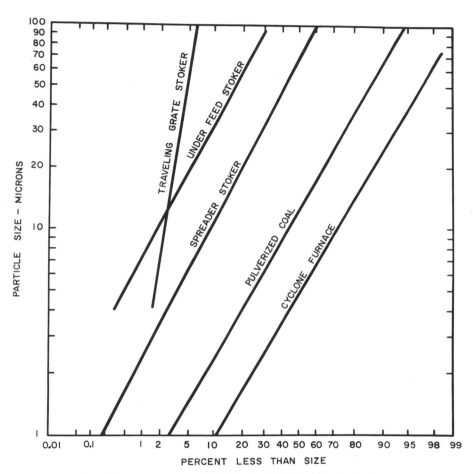

Fig. 9.2 Particle size distribution for different firing methods.

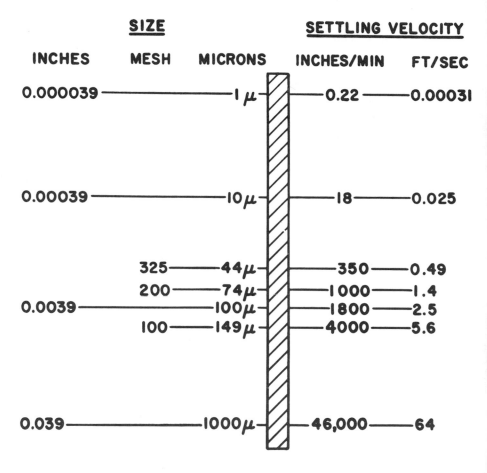

SIZE - SETTLING VELOCITY RELATIONSHIPS FOR SPHERICAL FLY ASH PARTICLES (SPECIFIC GRAVITY - 2.5)

Fig. 9.3 Size-setting velocity relationships for spherical fly ash particles (specific gravity: 2.5).

Possibly 80 percent of the nearly 50 million tons of particulate matter produced in 1970 from stationary sources was recovered by gas-cleaning devices. A wide variety of equipment is available for control of particulate emission from stationary combustion sources. The cost of equipment is usually greater for devices of high efficiency; however, the performance of competitively priced gas-cleaning equipment may differ considerably. Table 9.2 shows the optimum performance that may be expected from various types of gas-cleaning equipment used on stationary sources for removing particulate matter from flue gases.[3]

TABLE 9.2
Optimum Expected Performance of Various Types of Gas Cleaning Systems for Stationary Combustion Sources

	Removal of uncontrolled particulate emissions—%					
Sources	Settling Chambers	Large Diameter Cyclones	Small Diameter Cyclones	Electrostatic Precipitators	8-in. Pressure Drop Scrubbers	Fabric Filters
Coal-fired						
Spreader, chain grate, and vibrating stokers	50	60	85	99.5	99+	99.5
Other stokers	60	65	90	99.5	99+	99.5
Cyclone furnaces	10	15	70	99.5	†	†
Other pulverized coal units	20	30	80	99.5	99+	99.5
Oil-fired	5 *	10 *	30 *	75.0	†	†

SOURCE: *Control Techniques for Sulfur Oxides,* National Air Pollution Control Administration Publication AP-52 (January 1969).
* Efficiency estimated—not commonly used.
† Insufficient data for estimate.

Sulfur Oxide Emission

In 1968, approximately 33.2 million tons of sulfur oxide, primarily SO_2, was emitted in the United States. Figure 9.4 shows that nearly 76 percent, or 26 million tons, was produced by combustion of fossil fuels.[5] Of this amount, 16.8 million tons was emitted from electric utility boilers, 7.6 million tons by industrial boilers, and the remainder of the emissions from mobile sources. It has been estimated that the total SO_2 emissions will increase to 60 million tons by the year 2000 if no control measures are instituted (see fig. 9.5).[6] Of this total, it is estimated that 54 million tons would be from the combustion of fossil fuels.

In the combustion of fuels, the sulfur present is converted almost completely to sulfur oxide. The quantity of sulfur in fuels differs widely, ranging from less than 0.5 percent to more than 5 percent. Perry presents data

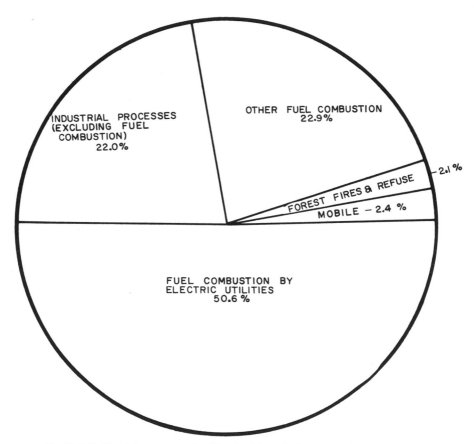

Fig. 9.4 Nationwide sources of sulfur dioxide emissions, 1968.
SOURCE: *Control Technique for Sulfur Dioxides,* United States Department of Health, Education, and Welfare, NAPCA Publication AP-52, January 1969.

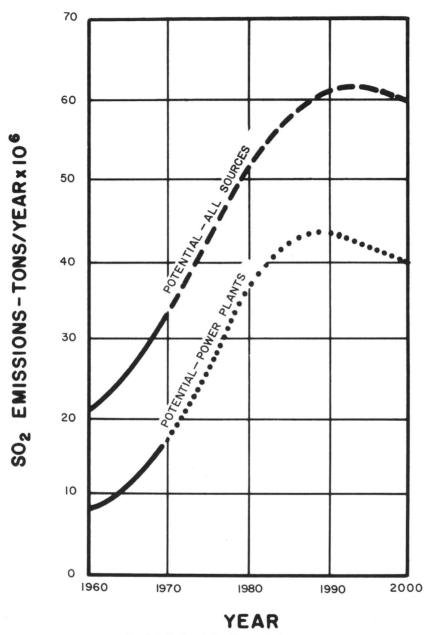

Fig. 9.5 Estimated sulfur emissions.

indicating that the average sulfur content of coal used for electric power production in the United States is approximately 2.5 percent.[7] Residual fuel oils may have a sulfur content from less than 1 percent to more than 4 percent, however, the fuel oils used for power production in the eastern United States commonly contain 2 to 3 percent. (Recent laws have reduced the level to one percent in many localities.) Natural gas usually contains very little sulfur. (If it contains appreciable amounts of H_2S, this is usually removed by the producer.)

In general, 90 percent or more of the sulfur in fuel will be present in the stack gases as sulfur oxides. Typically, concentrations will be 0.1 to 0.25 percent by volume. The degree of conversion of SO_2 to SO_3 is generally 1 to 4 percent. The formation of SO_3 in boilers is a complex process which is not thoroughly understood. Three possible mechanisms have been suggested: (1) oxidation of SO_2 by molecular oxygen; (2) oxidation of SO_2 in the flame by atomic oxygen; and (3) catalytic oxidation of SO_2.

Consideration of these mechanisms has been discussed by various investigators,[8-11] but extensive field tests on one boiler indicated that catalytic oxidation in the convection pass was the major source of SO_3.[12]

The control of emissions of sulfur dioxide is important because of the potential damage to vegetation and because of possible adverse health effects. Long-term exposure to low concentrations of SO_2 may contribute to chronic health problems, but the evidence on this point is not conclusive. Also, sulfur oxides contribute to corrosion and deposit problems in the boiler. Reductions in SO_2 emission can be accomplished by: (1) the use of low-sulfur fuel; (2) removal of sulfur from the fuel before burning; or (3) removal of SO_2 from the stack gas.

Low-sulfur fuels are in limited supply. Crude oils from North Africa, Nigeria, and the Far East are low in sulfur. Fuel oil produced from these crudes generally contains less than 1 percent sulfur, but the bulk of this oil does not reach the United States market. Most of our oil supply comes from the Southwestern United States or South America. Oil from these sources has a higher sulfur content, reaching as much as 5 percent in some cases.

Reserves of coal with less than 1 percent sulfur—more than one trillion tons—are ample. Of this huge reserve, however, about 80 percent (sub-bituminous or lignite) is found in the western part of the United States, a considerable distance from the large eastern coal-consuming centers. The remaining low-sulfur (bituminous) coal is reserved for metal production.

About one-fourth of the total known reserves of bituminous coal has a 2.2 percent sulfur content. These reserves, mostly located in eastern or midwestern states, are the main source of fuel for present day utility power plants.

Natural gas contains no sulfur and power plants fired with this fuel emit no SO_2 to the atmosphere. However, the supply of natural gas is limited, and with proven resources shrinking rapidly, the use of this premium fuel by utilities is decreasing. Although about 20 percent of the electricity generated in 1970 came from natural gas-fired plants, this figure is expected to drop to 15 percent by 1975 and 13 percent by 1980.

Removal of sulfur from fuel before burning would be an ideal solution to the problem of SO_2 emission, but available cost data for fuel desulfurization indicate that current methods are not economical. An extensive study of methods for desulfurizing residual fuel oil has been published by the United States Bureau of Mines.[13] The most promising method is hydro-desulfurization, where the oil is treated with hydrogen to form hydrogen sulfide gas which can then be separated from the liquid oil. Cost estimates for this process vary from 40 cents to one dollar per barrel, thus adding 25 to 50 percent to the cost of the oil. Similar costs have been estimated for selective blending or fractionation.

Generalized data on costs for sulfur removal from coal are even harder to obtain because of the wide differences in sulfur distribution among different coals. Organic sulfur, which comprises 50 to 60 percent of the total, is an integral part of the coal structure and cannot be removed by mechanical cleaning. Pyritic sulfur accounts for most of the remaining total sulfur. Finely divided pyritic sulfur, which is dispersed through the coal, is not easily removed by coal cleaning; but pyritic sulfur, which is found as large aggregates, can be removed effectively. Consequently, the extent of sulfur removal from coal is limited and somewhat variable, costing up to a dollar a ton, thereby adding 15 to 30 percent to the cost of the fuel.

Over the past five years, great interest has developed in cleaning flue gas to eliminate SO_2. More than twenty processes are under development.[14] Some of these methods are unsuited to power plants, but several are attractive. Six methods are getting major attention.

Historically, the first consists of simple water scrubbing of the flue gas to carry off SO_2 in solution. It was first used in 1932 at two large power plants in London. Costs and operating difficulties have discouraged additional installations.

Another appealing scheme is based on injecting pulverized limestone into the furnace. Gaseous SO_2 is then converted to solid calcium sulfate and carried away with the fly ash. However, dry injection of pulverized limestone or dolomite removes only 25 to 35 percent of the sulfur oxide [15] and the increased dust loading requires improved particulate matter collection. Work in Germany, Japan, and the United States has indicated limited success. A 3.3 million dollar, eighteen-month study by the Air Pollution Control Office of the Environmental Protection Agency at TVA's Shawnee

Power Plant is currently examining ways to increase the effectiveness of the dry injection procedure.

Wet absorption is by far the most popular system in terms of number of processes under development. At least a half dozen variations are under study. One leading process offered by Combustion Engineering employs furnace injection of an alkaline earth additive followed by wet scrubbing (see fig. 9.6). The pulverized additive calcines in the furnace reacting with combustion gases to form compounds of calcium and magnesium. This removes 20 to 30 percent of the sulfur oxides including all of the SO_3.

The flue gas containing unreactive SO_2 and calcined additive then passes into the wet scrubber. In the scrubber, the calcined additive that has not combined with the SO_2 in the furnace reacts with the water and the remaining SO_2 to form sulfates and sulfites of calcium and magnesium. At the same time, water entrainment of fly ash removes particulate matter. The solution containing the reacted materials drains out the bottom of the scrubber to a tank, clarifier, or pond where the particulates settle. Clarified water is then available for recirculation. The cleansed flue gas passes through a mist eliminator for removal of water and is then reheated for induced-draft fan protection.

Following successful pilot plant work at the St. Clair station of the Detroit Edison Company, C-E systems were installed at the 140-Mw. Meramec No. 2 unit of the Union Electric Company in St. Louis and the 125-Mw. Lawrence No. 4 unit of Kansas Power and Light Company. The third unit is installed at Kansas Power and Light Company on a new 430-Mw. boiler, with operation scheduled this year. Operations thus far have revealed problems not evident during pilot plant work. As a result, the Meramec and Lawrence systems have been modified in areas of additive injection, gas distribution, and water control.

In addition to removing more than 80 percent of the SO_2 from the flue gas, the systems also remove 99 percent of the particulate matter. Tail-end systems using calcium carbonate slurry without furnace injection are under development.

A basically different process converts SO_2 into SO_3 after the flue gas has been cleaned by taking out the particulate matter. The flue gas is then passed through a fixed catalyst bed. The sulfuric acid that results is collected in a condenser ahead of the stack. The process is being installed on a full-scale unit in a new midwestern power plant.

Still another approach is based on solvents capturing SO_2 by physical or chemical means. In this area, the alkalized alumina process of the United States Bureau of Mines has received the most attention. Alkalized alumina reacts with the SO_2 to form sulfates. Later regeneration releases the SO_2 and reforms the absorbant. Size degradation of the alkalized alumina during

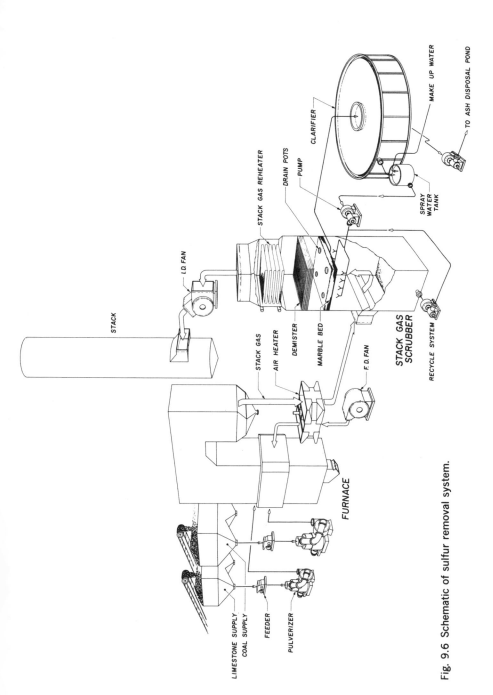

Fig. 9.6 Schematic of sulfur removal system.

177

repeated handling has presented serious problems and no large installations have been built.

A similar process, developed by Shell, uses a copper oxide catalyst. Pilot studies indicate that catalyst attrition is considerably less of a problem in this process.

Wet scrubbing systems that use chemical agents rather than water have been proposed by many investigators. One such process employs molten carbonates of lithium, sodium, and potassium to scrub the flue gas at about 1000° F. The resulting sulfates are reconverted later to carbonates and the sulfur is recovered. Power plant studies are being proposed, but no full-size installations are being installed.

With the increasing attention to air pollution, considerable effort has been placed on coal beneficiation to provide a low ash–low sulfur fuel. Major efforts appear to be in the area of gasification and solvent refining for de-ashed coal. In addition to improving air quality, such factors as fouling, slagging, and corrosion can be lessened. Work in this area is being done by a number of researchers.

Nitrogen Oxide Emission

Available data indicate that about 20 million tons of NO_x were emitted in the United States during 1968.[16,17] As shown in figure 9.7, approximately 87 percent of this total, or 17½ million tons, was from combustion of fossil fuels. Mobile sources accounted for 41 percent or nearly 8 million tons. Of the 36 percent, or 9 million tons emitted by stationary sources, power plants accounted for 3.6 million tons, industrial plants 2.8 million tons, and miscellaneous domestic sources 2.7 million tons. It has been estimated that the total NO_x emissions will double by the year 2000 if control measures are not instituted [17] (see fig. 9.8).

In contrast to sulfur oxides in combustion gases which are directly related to the sulfur content of the fuel, most of the NO_x formation and emission is the direct result of the reaction of nitrogen and oxygen gases present in the combustion air at high flame temperatures. In ordinary combustion calculations, it is customary to assume that the nitrogen in the combustion air is inert and does not react. From the standpoint of chemical activity, nitrogen is a relatively inert element; however, a mixture of nitrogen and oxygen subjected to the high furnace temperatures (2800 to 4000° F.) will react to produce a significant quantity of NO according to the reaction:

$$N_2 + O_2 \rightleftharpoons 2NO$$

For this reaction, thermodynamic considerations indicate that increased temperature favors the formation of NO provided chemical equilibrium is attained. However, in a practical situation, equilibrium conditions are not

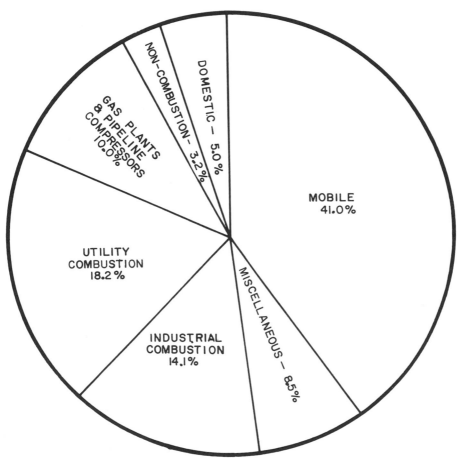

Fig. 9.7 Nationwide sources of nitrogen oxide emission, 1968.
SOURCES: "Air Pollution From Motor Vehicles," Battelle Research Outlook, vol. 2, no. 3, 1970.
W. Bartok et al., "Systems Study of Nitrogen Oxide Control Methods for Stationary Sources," NAPCA contract no. PH-22-68-55, November 1969.

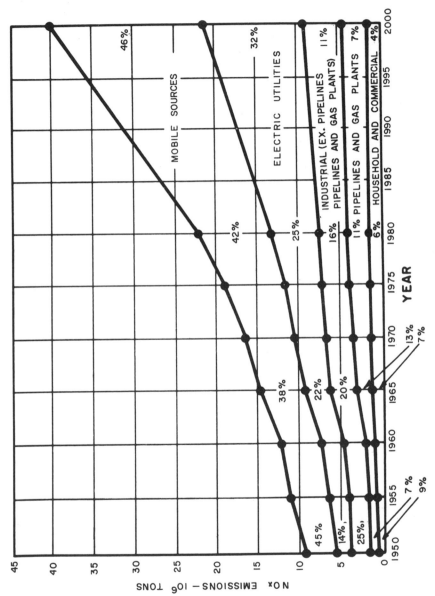

Fig. 9.8 Potential nitrogen oxide emissions in the United States.

reached and the rates of forward and reverse reactions must be taken into account. In actual practice less than 1 percent NO is formed. Spectroscopic studies have indicated that almost all the oxides of nitrogen in boiler flue gas are present as NO, with only a minor proportion present as NO_2.[18] It is the subsequent oxidation in the atmosphere of NO to NO_2 that is the source of a number of problems including the formation of photochemical smog and eye irritants, as well as the visible brown color when NO_x is emitted in sufficient concentration.

Because the formation process is quite involved, it is difficult to predict how much nitrogen oxide will be produced from a given furnace. The amount of NO_x may vary depending upon combustion conditions, the important ones being flame temperature, the time combustion gases are in the flame at high temperatures, and excess air present in the flame.

Table 9.3 indicates the significance of flame temperature on the production of NO_x.[19] With temperatures as low as 2000° F., 500 ppm. cannot be produced. At 2800° F. it takes sixteen seconds, whereas at 3600° F. it only takes one-ninth of a second to reach a concentration of 500 ppm. Since we are dealing in residence times of four to six seconds in modern furnaces, this means that temperatures above 2800° F. are the most significant. Limiting the maximum flame temperature to as low a value as practical is important; for example, control and gradual mixing of fuel and air in the furnace will slow the burning process and reduce the peak flame temperature. This is characteristic of the tangential firing method, wherein air is mixed with the fuel continually along the flame path resulting in complete combination of fuel and air, but reducing the peak flame temperature.

TABLE 9.3
Time for Nitrogen Oxide Formation at Elevated Temperatures

Temp, °F	Time to Form 500 ppm. NO, sec	NO Concentration at Equilibrium, ppm.
2000	—	180
2400	1370	550
2800	16.2	1380
3200	1.1	2600
3600	0.117	4150

SOURCE: J. A. Danielson, Air Pollution Engineering Manual, PHs Publication 999 AP-40 (1960).

Other procedures for reducing flame temperatures can also be important. If gas recirculation is properly introduced into the combustion process, it too can slow the rate of combustion and further prevent the flame from reaching adiabatic temperature, thereby minimizing the formation of oxides of nitrogen. Recent testing conducted in California indicates this quite dramatically. Reduction of NO_x from 250 ppm. to less than 100 ppm. has been achieved on oil-fired units. The use of two-stage combustion or over fire air is also effective in that both peak and average flame temperatures are reduced, and thus, NO_x formation can be minimized. Studies on tangential gas-fired units have shown that NO_x can be reduced from approximately 250 ppm. down to about 150 ppm. by variations in two-stage combustion. It has been reported that emissions from one horizontal gas-fired unit showed a reduction from 1200 to as low as 250 ppm. using two-stage combustion.[20]

To illustrate the effect of fuel and type of firing, reference is made to the following tabulation of data from more than twenty-five operating boilers:

Fuel	Firing	ppm. NO_x
Oil	Horizontal	500 to 700
Oil	Tangential	150 to 350
Gas	Horizontal	250 to 550
Gas	Tangential	50 to 300
Coal	Tangential	250 to 600

Note the markedly lower values for tangential firing for both gas and oil. The data, in general, agree with those of other investigators and indicate that NO_x emissions are highest for coal and decrease for oil and gas in that order.

Since NO is formed by the high temperature reaction of nitrogen and oxygen, one would expect that the amount of excess air available would affect NO production. Figures 9.9 and 9.10 show that a decrease in excess air produces a decrease in NO_x. The effect is particularly pronounced for horizontal firing (see fig. 9.9). Figure 9.10 shows the influence of varying excess air on three relatively new tangentially fired boilers using the three major fuels. In this figure, the excess air is decreased to a point approaching 2 percent. It can be seen that the NO_x values approach 100 ppm. on gas- and oil-fired units and are considerably reduced on coal-fired units.

NAVAJO GENERATING STATION

Measures taken to minimize emissions at the new Navajo Generating Station at Page, Arizona, lend perspective to this subject. Each of the three generating units at the Navajo Station will have a separate 750-Mw. coal-fired boiler. The average coal, which will be delivered to the Navajo Station, will have an ash content of approximately 8 percent and a heat content of 11,000 Btu. per pound.

Air quality in the area subject to the emissions from the Navajo Power Generating Station will be maintained by:

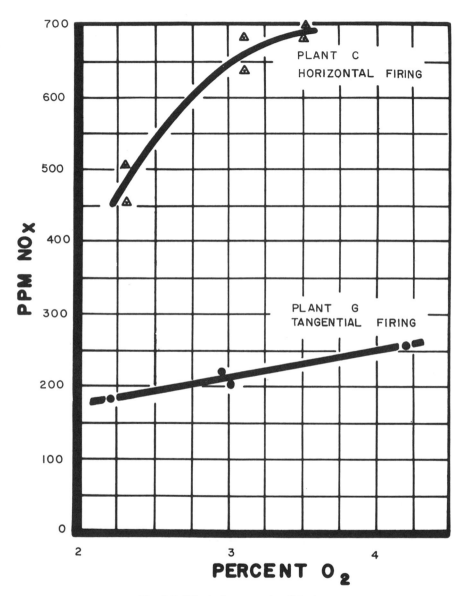

Fig. 9.9 Effect of excess air, oil fuel.

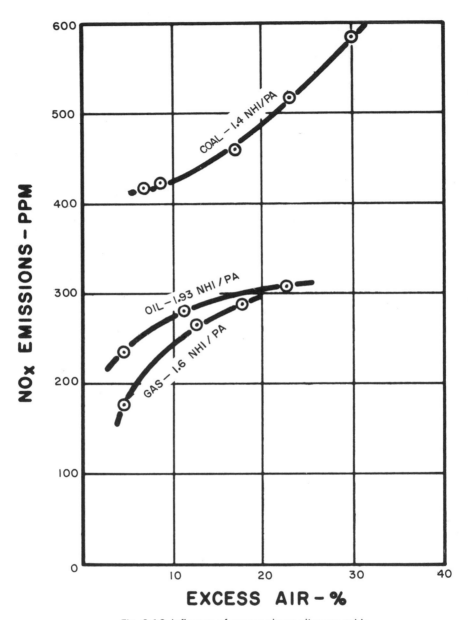

Fig. 9.10 Influence of excess air on nitrogen oxide.

1. Installing electrostatic precipitators or equivalent equipment for particulate removal with a design efficiency of 99.5 percent
2. Utilizing a low-sulfur coal (average 0.51 percent sulfur) and considering means of minimizing the maximum sulfur that can enter the furnace
3. Installing tall stacks to provide adequate dispersion of flue gases
4. Providing space for the later addition of a commercially proven SO_2 removal process
5. Installing a boiler which utilizes a tangential firing pattern and providing for later addition of other features which may reduce NO_x formation.[21]

In addition to tangential firing which is inherently beneficial in reducing NO_x, C-E will apply the latest features for improving the temperature-time relationship, such as alternate furnace air inlet locations, low excess air firing, flue gas recirculation, and over fire air.

SUMMARY OF EMISSIONS

Emissions from combustion sources that are significant from the standpoint of air pollution include particulate matter, sulfur oxides, and nitrogen oxides.

Particulate matter is objectionable on aesthetic grounds. The technology for its control is well developed, although efforts are constantly being made to improve collection equipment and reduce the cost.

Efforts to reduce emissions of sulfur oxides are prompted by their damaging effects on plant life and possible adverse health effects. From the equipment operator's standpoint, sulfur oxides can be detrimental in that they contribute to corrosion and deposit problems in the boiler. Techniques have been developed for control of SO_3 by means of low-excess air and/or additives. SO_2 control can be accomplished by use of low-sulfur fuel, fuel desulfurization, and removing the SO_2 from the stack gas. Progress has been made in utilization of all of these, but none is completely acceptable because numerous economic and technical engineering problems still persist. Work in this area is being actively pursued by many organizations throughout the world.

Oxides of nitrogen are air pollutants because of their participation in the reactions leading to photochemical smog. Since the localities most subject to photochemical smog are in oil and gas burning areas, most of the work has been done on these fuels. The emission of oxides of nitrogen can be significantly reduced by use of a suitable firing method to control the time-temperature relationship, low excess air firing, or an alternate fuel.

Reference notes

1. *Nationwide Inventory of Air Pollutant Emissions—1968,* National Air Pollution Control Administration Publication AP-73 (August 1970).
2. *The Tulsa Daily World,* Universal Science News, 7 March 1971.
3. *Control Techniques for Sulfur Oxides,* National Air Pollution Control Administration Publication AP-52 (January 1969).
4. J. D. Sensenbaugh and J. H. Fernandes, *Technical Orientation in Air Pollution* (Connecticut State Health Department, 1966).
5. *Control Techniques for Particulate Air Pollutants,* National Air Pollution Control Administration Publication AP-51 (January 1969).
6. F. A. Rohrman and J. H. Ludwig, *Power,* May 1967, p. 82.
7. H. Perry, *Proceedings of the American Power Conference,* vol. 27 (1965), p. 107.
8. G. Whittingham, *Transactions of the Faraday Society* 44 (1948): 141.
9. G. Whittingham, *Third Symposium on Combustion, Flame and Explosion Phenomena* (Baltimore, Md.: Williams and Wilkins, 1949): 453.
10. A. B. Hedley, *Fuel Society Journal* 13 (Sheffield University, 1962): 45.
11. A. Levy and E. L. Merryman, *Engineering for Power,* Translations ASME, Series A, 87, 116 (1965): 374.
12. J. T. Reese, J. Jonakin, and V. Z. Caracristi, *Combustion* 36, no. 5 (1964): 29.
13. H. C. Carpenter and H. C. Cottingham, *A Survey of Methods for Desulfurizing Residual Fuel Oil,* Bureau of Mines, Information Circular 8156.
14. J. Jonakin and P. Maurin, *Chemical Engineering* 77, no. 8 (April 1970).
15. *Sulfur Oxide Removal from Power Plant Stack Gas,* NAPCA Contract TV-29233A.
16. D. A. Trayser and F. A. Creswick, *Battelle Research Outlook* 2, no. 3 (1970).
17. W. Bartok, et al., *Systems Study of Nitrogen Oxide Methods for Stationary Sources,* NAPCA Contract PH 22-68-55 (November 1969).
18. A. J. Haagen-Smit, V. D. Taylor, and M. F. Brunelle, *International Journal Air Pollution* 2 (1959): 159.
19. J. A. Danielson, *Air Pollution Engineering Manual,* PHs Publication 999 AP-40 (1960): 540.
20. D. W. James, *Electrical World,* 1 February 1971, p. 46.
21. Bechtel Corporation, *Environmental Planning for the Navajo Generating Station,* September 1970.

10. Present and Future Airborne Emissions Control

William J. Moroz
Director, Center for Air Environment Studies, The Pennsylvania State University

Over the past few years a great deal has been thought and written about maintaining the quality of our environment. One is well aware that a major factor contributing to the deterioration of our environment has been the tremendous technological progress made in this country over the last fifty to one hundred years; progress that has provided ease of living, leisure time, mobility, and all of the other newly acquired comforts. These measures of technological progress have been coupled with a rapidly expanding economy and a rapidly expanding population, which the economists tell us is the sign of prosperity for all. Technology is criticized, however, for having paid too little attention to the important or the esthetic values of life.

Reviewing all that has been written and said over the past few years, there are several items that can be put down as fact and that provide a spring board for analysis of our environmental problems.

1. Our population is going to increase. We simply do not have the intelligence—or a strong enough moral incentive—to restrict population at the present time.
2. Most people desire more leisure, more ease, more material goods, more energy, more everything; if individuals do not want these things personally, they want them for their children. Whether *more* means *best* we do not know, but it has always meant *better* in the past.
3. Few, if any, individuals are mentally prepared for the personal, economic, and social disruptions which would result from a stable, nonexpanding economy.

If we continue as in the past, the consequence of all of these statements will be the environmental disaster predicted by the ecologists. The obvious conclusion is that we cannot continue as in the past. We must *plan, accept,* and *pay* for technological and social changes which will permit an improved standard of living with emphasis on *improved* environmental quality. I have absolute faith in our ability to do this—it is apparent that I am not a member of the "doomsday club"—so long as the desire to do so persists; it must be done by steady, continuous, long-term pressure because of the time lags involved and because any relaxation will yield a corresponding deterioration of the environment. The challenge to our society is simply to provide quality with quantity. This is going to require an input from young people from many disciplines with the energy and the time to bring uninhibited ideas to fruition.

EXAMPLE OF PRESENT EMISSIONS AND CONTROL

As an example of the sort of change that must be made, let us consider the power generation process. The power companies have been criticized for their waste heat releases and for their emissions of atmospheric pollutants. It should be pointed out clearly that they are not the only "bad-guys," but they have been singled out partly because they have kept very good records and they have made these records available. They also represent a major and rapidly expanding sector of our economy in response to *our* demands.

In terms of environmental control, there has been little real attention to the elimination of waste heat releases in the power industry. With reference to air pollution control, today any power station built will have highly efficient particle removal equipment, but no power station is yet completely equipped to control the sulfur or nitrogen oxides emissions. It is true that experiments to control sulfur oxides emission have been conducted reluctantly, and control devices are even more reluctantly considered for each new station built. It should also be noted here that some of the sulfur oxides control processes proposed also result in partial removal of nitrogen oxides. Unfortunately, the power companies feel that the proposed devices are not developed sufficiently well at this date, and no full scale installations have yet been made.

As a result of the nature of the example chosen, emissions of hydrocarbons or carbon monoxide are not included in this analysis—no self-respecting power station operator would permit significant emissions of these contaminants. In this respect, the analysis is more appropriate to air pollution problems of the cities of the Northeast than to cities on the West Coast of the United States. On the other hand, the concepts expressed here are generally applicable.

POWER CYCLE

In order to refresh our memory of the power generation cycle as it exists today, data typical of a modern power station are incorporated in figure 10.1. Technologically, we should note that only steam can be expanded in a turbine, hence a relatively large waste heat reject (latent heat of liquifaction) to the environment at the condensers. This cannot be avoided.

Focusing our attention on the emissions to the environment at the power plant, we note that for a modern 1580 Mw. power plant typical of current installations:

1. At the boiler there is an emission of approximately:
   a. 1.4×10^9 Btu. per hour to the atmosphere through the stack (excluding boiler radiation losses)
   b. 44,000 lbs. SO_2 per hour through the stack (assuming 2 percent S coal is being burned)
   c. 990 lbs. of small (<5 micron) particles per hour through the stack (assuming collectors of 99 percent efficiency and 9 percent ash coal)
   d. 11,800 lbs. NO_x per hour through the stack
2. At the condenser there is a thermal energy reject of 10^9 Btu. per hour to a water body or through cooling towers. If water is used this amount of energy can raise the temperature of a lake one mile long, one-eighth of a mile wide and ten feet deep by $10°$ F. each hour, neglecting evaporative and radiation losses. If cooling towers are used, 10.4×10^6 lbs. of water will be evaporated into the atmosphere each hour; this is the amount of liquid water contained in a fog bank of 7.28 micron droplets where visibility is reduced to one-tenth of a mile, having dimensions of 1 mile $\times$ 1 mile $\times$ 7.4 miles. While water vapor has not been traditionally viewed as an atmospheric contaminant, it can be, under the right circumstances and in the amounts released here.

With respect to thermal energy reject, you will note that all heat is rejected from the condensers at a temperature near that of the environment, i.e., at a sensible heat level which makes recovery economically and technologically impractical.

WASTE HEAT UTILIZATION

Let us modify the power cycle we have just described by installing a back pressure turbine to convert thermal to mechanical energy, instead of using a more conventional condensing turbine. This is relatively simple to do, as shown in the modified system diagram, figure 10.2, and would present no

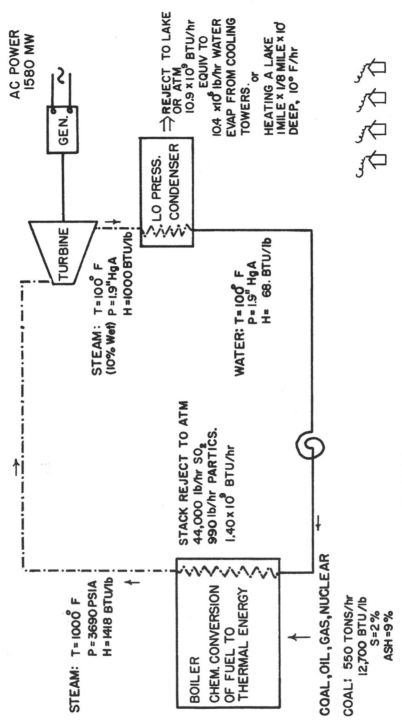

AC POWER 1580 MW

GEN.

TURBINE

STEAM: T=1000° F
P=3690 PSIA
H=1418 BTU/lb

STEAM: T=100° F P=1.9"HgA
(10% Wet) H=1000 BTU/lb

LO PRESS. CONDENSER

→ REJECT TO LAKE
OR ATM
10.9 x 10⁹ BTU/hr
EQUIV TO
10.4 x10⁶ lb/hr WATER
EVAP FROM COOLING
TOWERS.
or
HEATING A LAKE
1MILE x 1/8 MILE x 10¹
DEEP, 10⁹ °F/hr

WATER: T=100° F
P=1.9"HgA
H= 68. BTU/lb

STACK REJECT TO ATM
44,000 lb/hr SO₂
990 lb/hr PARTICS.
1.40 x 10⁹ BTU/hr

BOILER
CHEM. CONVERSION
OF FUEL TO
THERMAL ENERGY

COAL, OIL, GAS, NUCLEAR

COAL: 550 TONS/hr
12,700 BTU/lb
S=2%
ASH =9%

Fig. 10.1 Typical parameters for a modern power system.

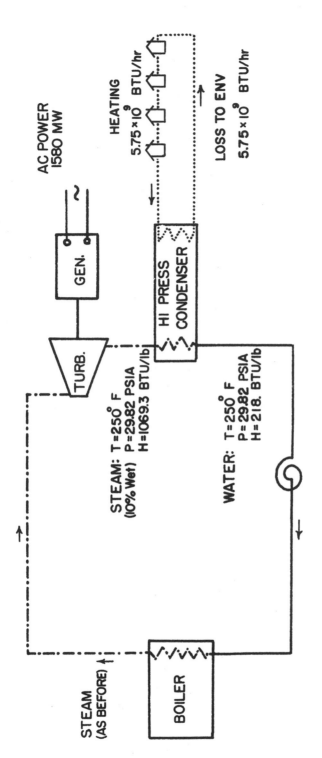

Fig. 10.2 Systems parameters for a waste heat utilization power system.

191

problems. For simplicity we will consider a turbine exhausting at about 30 psia. At this pressure the exhaust steam temperature would be 250° F., and thermal energy could be recovered to heat water in a simple heat exchanger similar to the condenser. The hot water could then be used to heat apartments, stores, office buildings, houses, greenhouses, etc., in winter, or the steam could be used directly for air conditioning, using an absorption refrigeration cycle, in summer. In spring and fall the energy might be dumped, at least in part, into a man-made recreation lake so the swimming and recreational period would be extended. Alternatively the energy could be used in a desalinization facility or in a paper plant for drying. At worst we could go back to our present systems during these periods.

It is noted that under the proposed cycle we would reject more thermal energy to the environment from the power plant, and that the rejection temperature would be considerably higher. Let us look at the effect of this action.

As a result of the greater thermal energy rejection, the original power plant would now generate only 1355 Mw. of electrical energy. In order to increase the electrical capacity of the power plant, it would have to burn more fuel and there would be correspondingly increased emissions as follows:

1. At the boiler the new figures (assuming coal is still used as fuel):
   a. 1.66×10^9 Btu. per hour thermal energy rejected to the atmosphere through the stack
   b. 52,000 lbs. SO_2 per hour would be emitted through the stack
   c. 1180 lbs. per hour of small particles would be emitted through the stack
   d. 14,000 lbs. NO_x per hour emitted through the stack
2. At the turbine exhaust there is a thermal energy rejection of 11.5×10^9 Btu. per hour, but this is at 250° F. and 30 psia. Most of this energy can be recovered in a heat exchanger where condensation takes place at the higher temperature.

For purposes of our analysis we will consider that water is heated in the high pressure condenser and that this water is used to heat domestic residences or apartments. The hot water will have to be distributed through a pipeline system over a very large area and it is found that only 50 percent of the energy can actually be recovered for heating; the balance is lost, primarily in transmission. Again, for analysis purposes, we will assume that the residences to be heated are presently using #2 fuel oil at 75 percent efficiency (a high average figure), and that they have an average heat demand of 50,000 Btu. per hour.

As a result of losses from the hot water distribution system, of the

11.5 X 10⁹ Btu. per hour recovered in the high pressure condenser, only 50 percent, or 5.75 X 10⁹ Btu. per hour, is available for residence heating. This is the amount of energy which would be put into the residences from burning 298 tons of coal per hour, or 54,500 gallons of fuel oil per hour, or 7.74 X 10⁶ cu. ft. of natural gas per hour.

If the residences were burning oil of 0.5 percent sulfur content and 0.1 percent ash content, the emissions from the residence chimneys would total:

1. 4360 lbs. of SO_2 per hour
2. 436 lbs. of particles per hour, neglecting any carbon or soot emission
3. 1.44 X 10⁹ Btu. per hour direct thermal discharge from the residence chimneys *
4. 2400 lbs. per hour of NO_x emitted through residence chimneys

If the oil heating systems were replaced by a central heating system using the hot water generated at the condenser coupled to the back-pressure turbine, all of these emissions would disappear; a new net emissions balance (which includes the 50 percent of the energy lost from the hot water distribution system) is given in table 10.1.

We have done very well reducing both the waste heat rejection and the particle emissions in the community. Furthermore, if we review the literature on SO_2 recovery we find that by the simple addition of dolomite (limestone) in the furnaces of the power generating station we can achieve adsorption of about 30 percent of the SO_2 in the flue gases. As a consequence of the addition of solids at this point and the modification of exhaust gases, it will be necessary to redesign and enlarge the dust collectors of the plant with a corresponding cost increase. This additional cost must be offset by the financial return from the sale of hot water to residence owners. With dolomite-firing equipment installed, the new SO_2 emissions will be 36,400 lbs. per hour, which is well below the former emission. We have used no special equipment in this operation nor any technology which does not already exist. Alternatively, the fuel oil formerly used in the residences could be fired in the power plant replacing some of the coal required, and the SO_2 emissions would decrease to 32,360 lbs. per hour. However, it is felt that this is wasting a natural resource.

Alternatively, let us go back to the case where no SO_2 control was installed. All of the emissions from the new plant complex will be discharged from the tall stack at the power plant, and there will be no emissions from thirty foot residential chimneys. We know that the maximum contaminant concentration at the surface varies approximately inversely as the square of the height of emission. This means that a slightly taller stack

---

* Ultimately, all of the energy in the fuel is lost into the atmosphere from the houses.

TABLE 10.1
Comparison of Oil and Central Heating Systems

	Old System *	New System *
Net SO_2 emitted	$44,000 + 4,360 \doteq 48,360 \#/hr.$	52,000 #/hr. (using coal only)
Waste heat reject —stacks	$1.4 \times 10^9 + 1.4 \times 10^9 = 2.8 \times 10^9$ Btu./hr.	1.7×10^9 Btu./hr.
Waste heat reject —condenser	10.9×10^9 Btu./hr.	5.8×10^9 Btu.hr. (transmission losses)
total	13.7×10^9 Btu./hr.	7.5×10^9 Btu./hr.
Particles	$990 + 436 = 1370 \#/hr.$	1180 #/hr.
Nitrogen oxides	$11,800 + 2,400 = 14,200 \#/hr.$	14,000 #/hr.
Fuel savings		Savings of 54,500 gals. fuel oil/hr., but increased consumption of 100 t/hr. coal.

* The System referred to here is the combined power plant, residence heating system. The power plant for both the old and the new system is fired by coal. To reduce the SO_2 emissions from the New System we can: (1) use dry dolomite firing for SO_2 removal (about 30 percent removal), 36,400 #/hr. SO_2 emitted; and (2) use the fuel oil from the residences in the power plant, 32,560 #/hr. SO_2 emitted with savings of 298 tons/hr. of coal. To reduce the NO_x emission from the New System: Both (1) and (2) above will reduce NO_x emission, but it is difficult to estimate the amount of reduction due to inadequate data at this time.

could be installed at the power plant for the new system and surface concentrations of SO_2 would not increase, despite increased emissions from the plant even without controls. Experimental numerical modelling of pollution dispersion, presently being conducted at Pennsylvania State University, clearly shows this effect in long-term averages. On the other hand, dispersion is no solution and one would prefer to see the SO_2 removal process installed.

DISADVANTAGES OF THE NEW SYSTEM

It is apparent that the new system would have to be designed and operated as a unit—the system concept must be employed. With so many advantages why is this system not used extensively today? The Russians are using it as they have more central heating systems in use than all other countries of the world combined, but this is not an adequate answer.

There are complications, a few of which we can mention:

1. The power plant must be located near the thermal energy consumer; in this case, in the heart of a large city, for it would heat about 115,000 residences.

2. Discharges during the spring and fall are problematical when no heating or cooling is required.
3. It is doubtful that costs would be reduced because of the high cost of the heat distribution system.
4. It could be used only in the north where the heat is required or in the south where year-round air conditioning is required.

On the other hand, by all standards, we have reduced contamination of our environment, yet the availability of goods, energy, and services has not been reduced; in fact, it is suggested that availability of these commodities is improved, for there are no more domestic furnaces for the public to maintain. Consider the convenience.

THE FUTURE: REUSE, RECYCLING, AND RECOVERY

We have only discussed a single industry in our example, but there are an infinite number of examples which one could bring to the attention of others. It will take all of us to handle the job of environmental improvement.

Thermal pollution, or thermal enrichment as it is sometimes called, can be expected to increase linearly with the demand for energy in all forms as step-function decreases of waste energy dissipation requirements (alternatively, increases of process efficiency) are not probable in the immediate future. Both dry- and wet-type waste heat dissipation devices have been designed for stationary equipment; both dissipate waste to a local water body or to the atmosphere in amounts which may be limited by other environmental considerations. Again, considering an expanding population and a simultaneously expanding demand, problems associated with waste energy dissipation will become more frequent and more acute unless methods for recycling or recovery are adopted.

Looking beyond the individual problems it is apparent that the problems, like the control techniques, are interrelated. If waste energy can be utilized, the immediate consequence is a reduction of thermal pollution. At the same time, less fuel will be required with the result that contaminant emission to our atmosphere will also be reduced. Higher energy utilization factors will force us to design with more detail and will permit the specification of more sophisticated and expensive recovery systems. Foresight is always poor, but it is this author's opinion that we must move in this direction for we cannot achieve large improvements by any other method—improvements must be made more rapidly than we have been making them in the past. Reuse, recycling, and recovery seem to be our only alternatives.

Changes of the type proposed will require appreciable changes in social attitudes for both scientists and technologists and for the public. Priority changes are also desirable to conserve our available resources. The current public awareness of and emphasis on the quality of our environment may be replaced by other priorities in a few years; it is to be hoped that scientific awareness will not wane, for science has the responsibility of providing the leadership to proceed to a solution. The solution will require substantial scientific and technological advancement which could lead to the rebirth of our cities; and a rebirth is necessary for one cannot mess up the house and move out—there are limited places to move and our problems seem to follow us.

QUEST FOR ENVIRONMENTAL CONTROL

Limited aspects of environmental control have been touched upon in this chapter. In summary, one might refocus attention on the engineering approach to environmental control. We are making reductions in emissions of all types, but our gains on individual sources are more than offset by an increasing population and increasing per capita demand for goods, energy, and services. We may make temporary advancement at individual locations, many of which are at the expense of more rapid utilization of more valuable resources, but at present rates our ultimate advances will be backward. We must rely more heavily on reuse, recycling, recovery and on total energy utilization if significant advances are to be made in our quest for environmental control.

11. Thermal Emissions Control

D. G. Daniels and J. R. Eliason
Battelle-Northwest Institute

The number of power plants operating along rivers, estuaries, coastal areas, and on inland lakes is increasing, and projections based on the rate of increase suggest that many additional plants soon will have to be located in order to meet the growing power demands of the future. Increased demand for power requires sound advanced planning to minimize adverse effects upon the environment. As a result, predictive techniques are necessary to estimate, in advance, the environmental effects of these plants. Using these estimates, improved methods of plant siting and design can be evaluated.

A water transport simulation system has been developed by Battelle-Northwest; the system predicts temperature or concentration fields that would exist in the vicinity of an industrial effluent discharge. Lateral and longitudinal eddy diffusivities are calculated from dye studies, using remotely sensed data acquired from an aircraft and reduced by computer. A program has been written to solve the equations of motion in simplified form. The output from this program is the predicted current velocity field in the vicinity of the plant effluent discharge outside the turbulent mixing zone caused by momentum and buoyancy of the effluent jet. Dye velocity measurements taken during the dye studies are compared with the predicted velocity field for verification.

Within the turbulent mixing zone a separate jet model is used to predict the temperature or concentration and velocity fields. These temperatures or concentrations and velocities are used as boundary conditions in the simplified equations of motion and in the transport model. Final output is in the form of predicted isothermal or isoconcentration plots that can be

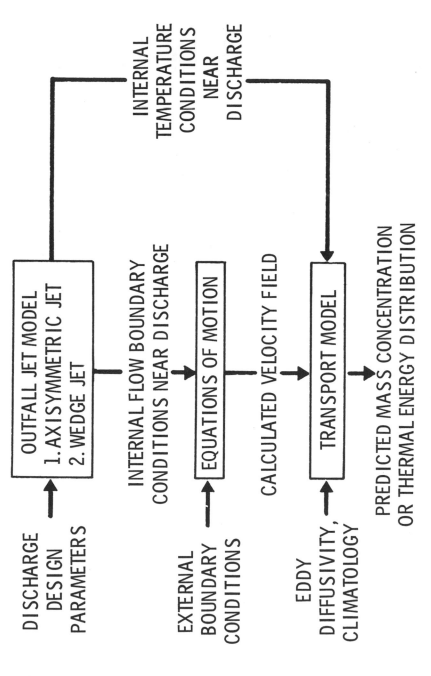

Fig. 11.1 Water transport simulation system which predicts temperature or concentration patterns.

used for advanced planning and determination of the effects of the industrial discharge on the environment. Also, temperature or concentration surfaces can be plotted for qualitative evaluation.

WATER TRANSPORT SIMULATION SYSTEM

The water transport simulation system is used to predict the temperature or concentration patterns that would exist in the vicinity of an industrial discharge and is based on the concept depicted in figure 11.1. Modeling the temperature or concentration and flow distribution in the region affected by effluent jet momentum and buoyancy is based on an outfall jet plume model developed by Baumgartner and Trent.[1] The velocity boundary conditions from this model together with external boundary conditions, ambient currents, and cooling water flow are input to a program solving simplified equations of motion, which predicts the flow field in the region affected by a discharge. The velocity field calculated by the equations of motion is input to the transport model together with diffusion coefficients calculated from remotely sensed dye data and climatology information. This transport model predicts the temperature or concentration field based on input conditions.

Outfall Jet Model

The outfall jet model simulates the transport of effluent from the discharge through the region where jet momentum and buoyancy are significant. There is considerable rotation, mixing, and buoyancy in this region, and equations of motion must account for these phenomena.

In this region, the temperature, mass concentration, and velocity fields are modeled with a two-part outfall model, the results of which are entered as boundary conditions into a simplified set of equations of motion. This two-part outfall model is composed of an initial mixing model used to transport the effluent water from the discharge to the surface or to the elevation where the mixed jet density is the same as the ambient water. From this point downstream, the plume is treated as a horizontal wedge.

The initial mixing model is based on a similarity solution of momentum, buoyancy, and dilution conservation equations together with geometric relationships for a horizontally-discharged buoyant jet. Figure 11.2 illustrates this model.

The wedge jet is carried beyond the location where the axially symmetric jet reaches its equilibrium elevation until the momentum in the wedge jet is about equal to the momentum of the receiving body. The velocities around this wedge and the temperatures calculated within this wedge jet are used as boundary conditions in the equation of motion and in the transport model.

Beyond this initial mixing model, the water is treated as a horizontal wedge moving at the elevation of its equilibrium density.

Equations of Motion

Velocity information necessary to predict the mass or energy transport is calculated from a simplified set of equations of motion. At present, these simplifications are based on the assumptions of: (a) steady flow; (b) coriolis forces being negligible; (c) buoyancy forces being negligible; (d) frictional forces being negligible; and (e) two-dimensional lateral irrotational flow. With these assumptions, the equations of motion can be written as follows:

$$u \frac{\partial u}{\partial x} + v \frac{\partial u}{\partial y} = - \frac{1}{\rho_0} \frac{\partial P}{\partial x} \tag{1}$$

$$u \frac{\partial v}{\partial x} + v \frac{\partial v}{\partial y} = - \frac{1}{\rho_0} \frac{\partial P}{\partial y} \tag{2}$$

It can be shown that Equations (1) and (2) are satisfied by

$$\nabla^2 \phi = 0 \tag{3}$$

where ϕ is a scalar potential given by

$$u_i = \frac{\partial \phi}{\partial x_i} \tag{4}$$

Dye studies, conducted in the surface waters of interest, yield field velocity data. Velocities for the dye plume are calculated by measuring plume movement with time according to the following equations:

$$u = \frac{x_2 - x_1}{t_2 - t_1} \tag{5}$$

$$v = \frac{y_2 - y_1}{t_2 - t_1} \tag{6}$$

A diagram of this relationship is given in figure 11.3. The x and y coordinates are measured from an arbitrary origin and represent the point in the dye plume with the maximum concentration. These measured velocities are compared with calculated velocities, and boundary potentials are adjusted until a reasonable comparison is obtained.

Transport Model

The transport model is based on the mass or energy conservation equation described conceptually in figure 11.4. This equation states that the rate of mass or energy entering a system, less the rate of mass or energy leaving a system, plus the amount of mass or energy created in the system, must equal the rate of accumulation of mass or energy within the system.

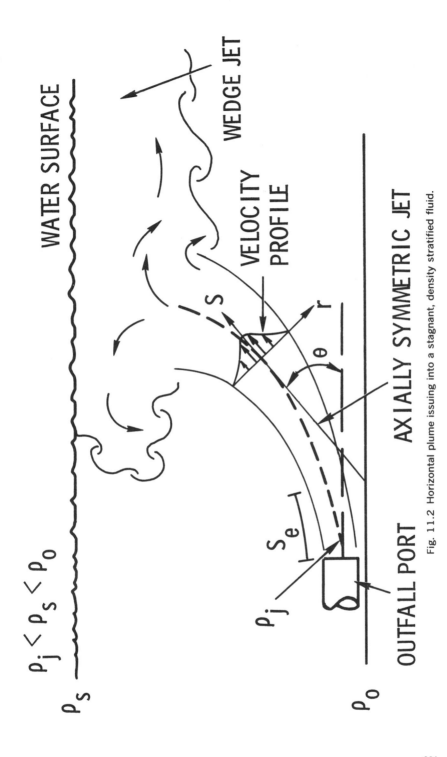

Fig. 11.2 Horizontal plume issuing into a stagnant, density stratified fluid.

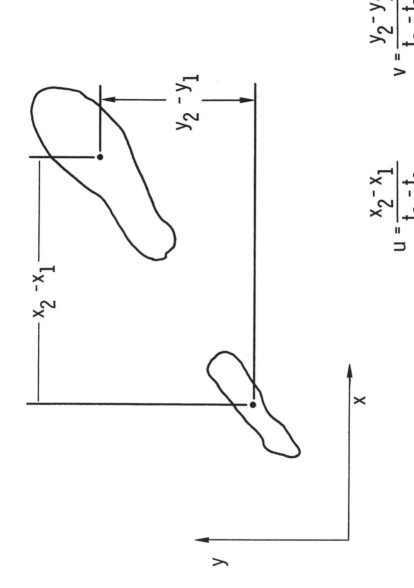

$$u = \frac{x_2 - x_1}{t_2 - t_1} \qquad v = \frac{y_2 - y_1}{t_2 - t_1}$$

Fig. 11.3 Velocity determination from dye studies.

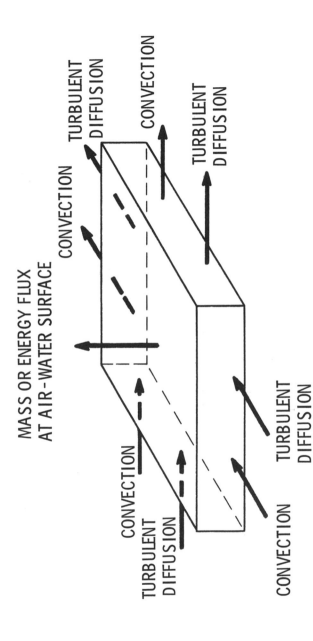

MASS OR ENERGY FLUX
AT AIR-WATER SURFACE

RATE OF ACCUMULATION = RATE IN - RATE OUT + SOURCE TERMS

$$\frac{\partial S}{\partial t} = -u\frac{\partial S}{\partial x} - v\frac{\partial S}{\partial y} + D_x\frac{\partial^2 S}{\partial x^2} + D_y\frac{\partial^2 S}{\partial y^2} + S_T$$

Fig. 11.4 Mass of energy balance on a cell.

Now if the system is treated as a block as shown in figure 11.4, the conservation equation can be written. When this block is reduced to infinitesimal size the conservation equation becomes

$$\frac{\partial S}{\partial t} = -u \frac{\partial S}{\partial x} - v \frac{\partial S}{\partial y} + D_x \frac{\partial^2 S}{\partial x^2} + D_y \frac{\partial^2 S}{\partial y^2} + S_T \qquad (7)$$

which includes convective, diffusive, and source terms for a two-dimensional, transient system with variable velocity field. Equation (7) cannot be solved analytically for a variable velocity field, but has been solved numerically by writing it in finite difference form using an alternating direction implicit algorithm. A detailed description of this algorithm is given in a recent publication by Oster, Sonnichsen, and Jaske.[2]

The system to be modeled is divided into a set of cells by a rectilinear grid system, and the finite difference algorithm is applied to each cell. The assumptions necessary to write Equation (7) in finite difference form include: (a) velocity being constant across a cell face; (b) eddy diffusivity being constant within the system; (c) all internal source or sink terms can be described by S_T; and (d) turbulent diffusion being treated analogous to microscopic diffusion but on a larger scale.

Determination of Eddy Diffusivity for Input to Transport Model

In order to solve Equation (7), eddy diffusivity and velocity must be known for the system. The velocity field may vary in time and space; the diffusivity field is assumed to be constant with respect to time and space. Lateral and longitudinal eddy diffusivity data are determined from remotely-sensed dye tracer data obtained with instrument systems developed at Battelle-Northwest. (See appendix to this chapter for a brief description of these systems.) This is done by assuming the dye plume is vertically homogeneous and that the concentration patterns follow a bivariate normal distribution in the lateral and longitudinal directions. The second assumption implies that the dye disperses according to the conservation equation described by Equation (7). If the velocity components and eddy diffusivities are assumed constant with respect to spatial coordinates and time, the conservation equation has an analytical solution

$$S = \frac{W_d}{2\pi\sigma_x\sigma_y} \exp\left\{-\tfrac{1}{2}\left[\left(\frac{x-ut}{\sigma_x}\right)^2 + \left(\frac{y-vt}{\sigma_y}\right)^2\right]\right\} \qquad (8)$$

where
$$\sigma_x = \sqrt{2\,D_x t} \qquad (9)$$

$$\sigma_y = \sqrt{2\,D_y t} \qquad (10)$$

and S refers to dye concentration.

The eddy diffusivities can be calculated by examining changes in the dye plume concentration as it moves and disperses with time. For example, by following a dye plume and obtaining successive detailed dye concentration patterns in time * with the remote sensing instruments, the effective average eddy diffusivities responsible for dispersal of the dye plume between successive time planes can be calculated. If the time difference between successive time planes is short and the dye plume does not extend over too large an area, the assumptions necessary to write Equation (8) are nearly satisfied.

The eddy diffusivities can be calculated from variance data obtained between adjacent time planes as

$$D_x = \frac{\sigma_{x_2}^2 - \sigma_{x_1}^2}{2(t_2 - t_1)} \tag{11}$$

$$D_y = \frac{\sigma_{y_2}^2 - \sigma_{y_1}^2}{2(t_2 - t_1)} \tag{12}$$

where subscripts 1 and 2 refer to variances calculated at the first and second time planes, respectively.

The evaluation of the lateral and longitudinal variances follows from the development of Equation (8) by Diachishin in 1963.[3] By examining the isoconcentration lines within the dye plume and measuring the surface areas of the regions enclosed within each isoconcentration line, one can plot the area versus the logarithm of the corresponding concentration line enclosing the area, as shown in figure 11.5. Examination of the table accompanying each remotely sensed dye concentration pattern (as in fig. 11.6) will show that by summing the right-hand column (which tabulates the area of each separate concentration range) and noting the lower limit of each concentration range, the above information is immediately available.

Diachishin has shown that each graph of the above described data should have a slope of $-2 \, \pi \sigma_x \sigma_y$. For example, a plot of some of these data is shown in figure 11.7. With further information regarding the ratio of σ_x and σ_y, which is obtained from the overall length and width of the dye plume, it is possible to calculate σ_x and σ_y for each time plane represented. From these variances, longitudinal and lateral eddy diffusivities can be calculated between successive time planes using Equations (11) and (12).

Generally accepted theories of transport phenomena in turbulent systems (e.g., the Prandtl mixing length theory) conclude that values of eddy mass diffusivity and eddy thermal diffusivity are essentially the same for systems in which the molecular rate of transport of mass and heat are comparable in magnitude. This is true for water and, consequently, the eddy mass

---

* Hereafter referred to as time planes.

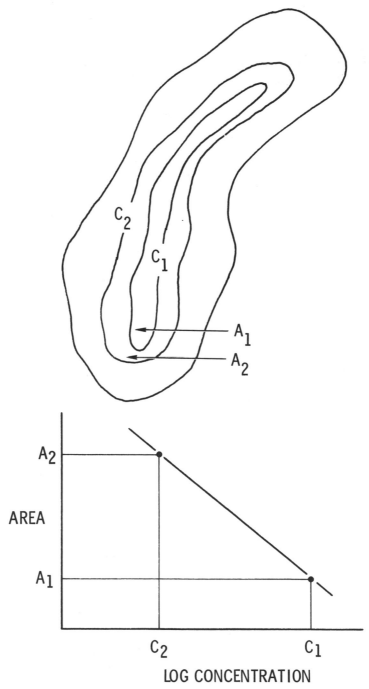

Fig. 11.5 Idealized dye concentration pattern.

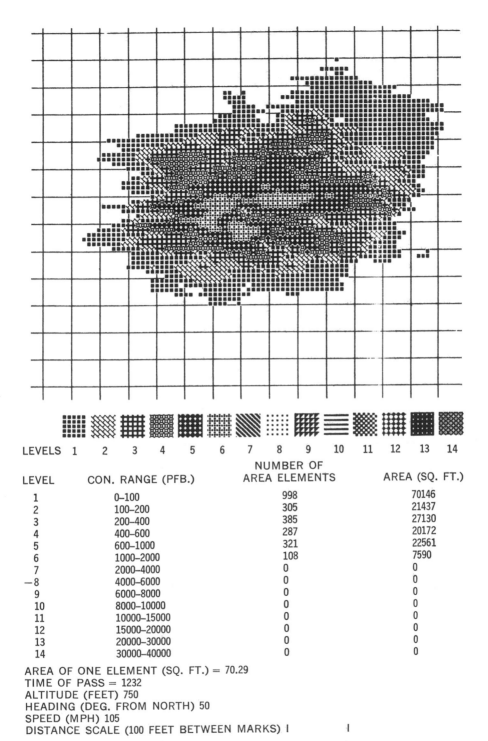

LEVELS 1 2 3 4 5 6 7 8 9 10 11 12 13 14

LEVEL	CON. RANGE (PFB.)	NUMBER OF AREA ELEMENTS	AREA (SQ. FT.)
1	0–100	998	70146
2	100–200	305	21437
3	200–400	385	27130
4	400–600	287	20172
5	600–1000	321	22561
6	1000–2000	108	7590
7	2000–4000	0	0
−8	4000–6000	0	0
9	6000–8000	0	0
10	8000–10000	0	0
11	10000–15000	0	0
12	15000–20000	0	0
13	20000–30000	0	0
14	30000–40000	0	0

AREA OF ONE ELEMENT (SQ. FT.) = 70.29
TIME OF PASS = 1232
ALTITUDE (FEET) 750
HEADING (DEG. FROM NORTH) 50
SPEED (MPH) 105
DISTANCE SCALE (100 FEET BETWEEN MARKS) I   I

Fig. 11.6 Computer generated tracer dye analysis.

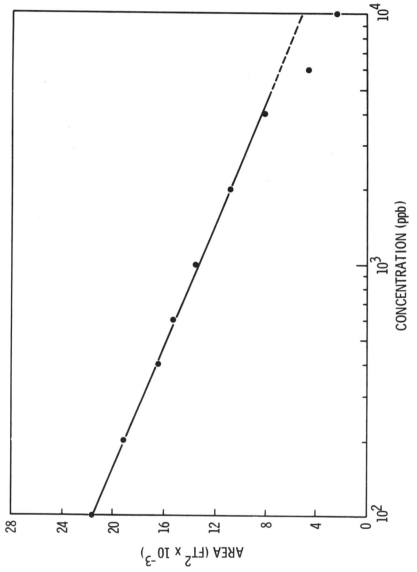

Fig. 11.7 Dye plume surface area as a function of concentration.

208

diffusivities derived from dye data are used directly in solving Equation (7) for mass concentration or thermal energy distribution.

Water Transport Simulation System Output

Using the eddy diffusivities calculated from the remotely sensed dye concentration data and the velocities calculated with the simplified equations of motion and source terms as input to the finite difference form of Equation (7), mass concentrations or temperature can be calculated for a particular problem.

An example is a simulation of the temperature distribution around a proposed industrial site. Figure 11.8 describes the potential contours calculated around the outfall of this plant located on the rectangular island. The streamlines run normal to the potential contours and in the direction of increasing potential. Figure 11.9 is a description of the temperature excess isotherms calculated for this case.

Figure 11.10 is a two-dimensional projection of the three-dimensional temperature excess surface described by temperature contour lines in figure 11.9. To give some idea of the scale in figure 11.10, the long side of the island is about 2800 feet long and the temperature spike near the center of the depicted surface is 3.5° C.

TECHNOLOGICAL DEVELOPMENTS

The predictive modeling techniques have been developed by Battelle-Northwest to simulate mass or energy transport in surface waters using two-dimensional finite difference equations with appropriate similarity assumptions in the third dimension. Simplified equations of motion have been developed to model the velocity field. This velocity field, together with eddy diffusivities obtained from remotely sensed dye concentration data from dye studies and appropriate boundary conditions, is input to a transport model capable of modeling the transport of industrial effluent in surface waters.

The direct use of data acquired with the remote sensing instrument systems, in conjunction with the predictive water transport simulation system to simulate the water effects of an effluent discharge on a body of water, has been accomplished for the first time in the comprehensive system described herein. More advanced methods of utilizing the remotely sensed data and of modeling the water systems are under development.

NOMENCLATURE

S Temperature or concentration field, °C. or ppm., respectively

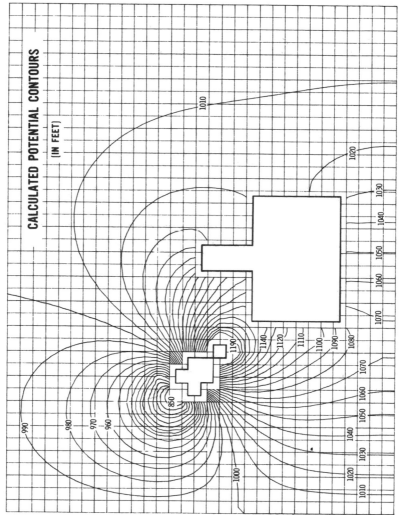

CALCULATED POTENTIAL CONTOURS

(IN FEET)

Fig. 11.8 Calculated potential contours for a discharge and intake near an island.

t	Time, seconds
D_x, D_y	Eddy mass and eddy thermal diffusivity in the x and y directions, respectively, ft.²/sec.
σ_x, σ_y	Variances in the x and y directions, respectively, ft.
W_d	Dye release parameter, ppm./ft.²
S	Distance along jet centerline where not referring to temperature or concentration field, ft.
x, y, z	Position coordinates directed east, north, and vertically upwards from an arbitrary origin
u, v, w or u_1, u_2, u_3	Velocity components in the x, y, and z directions, respectively, ft./sec.

APPENDIX

Remote sensing systems developed by Battelle-Northwest are now capable of providing diffusion data on tracer dyes for use in predictive models of surface water transport patterns.

The basic data collection system is an optical mechanical scanner which is operated from a light aircraft. Data is recorded directly on magnetic tape which can be input to a computer system for analysis. Output can be obtained as isothermal or isoconcentration plots produced directly by the computer, or the data can be output on digital tape, paper tape, or computer cards for subsequent computer analysis.

The optical mechanical imaging systems scan an area normal to the aircraft flight path up to sixty degrees either side of nadir. The lower right hand corner of figure 11.11 illustrates the relationships of the scanning system to the water surface for remote sensing of the surface concentrations in a dye plume. The dye is excited by sunlight, and the imaging system senses radiation emitted by the dye in a narrow bandpass around the peak emittance intensity for the dye. A similar scanning system is used to map surface water temperatures within a thermal plume by sensing long wave radiation emitted by the thermal plume. These data can be used to refine predictive modeling techniques or for thermal emissions control.

The emitted intensity (fluorescence or radiant energy) is converted into an electrical signal by a detector (photomultiplier or infrared detector) producing a signal that varies in amplitude with variations in the emitted intensity. A preamplifier and amplifier increase the output voltage of the detector, and the resulting signal is stored on magnetic tape. The scanning unit consists of a rotating mirror system that scans a segment of the surface normal to the flight path, and an optics system that collects and focuses the incoming radiant energy on the detector.

Since the radiant energy is a well-defined function of dye concentration

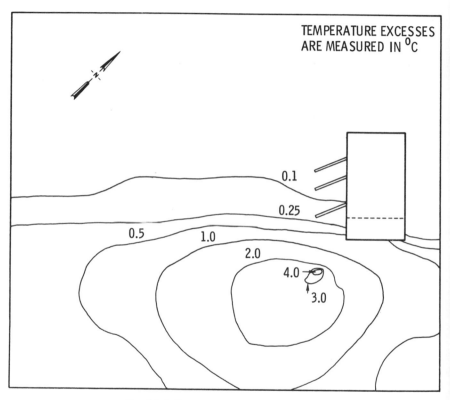

Fig. 11.9 Temperature excess isotherms.

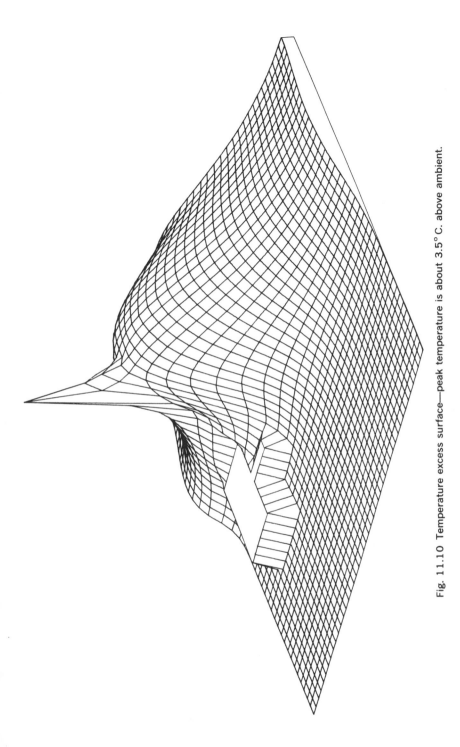

Fig. 11.10 Temperature excess surface—peak temperature is about 3.5° C. above ambient.

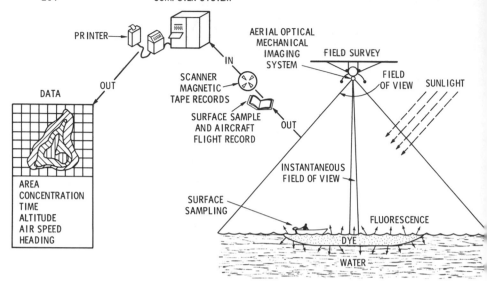

Fig. 11.11 Aerial dye survey and analysis system.

Fig. 11.12 Computer processed infrared temperature data.

LEVEL	TEMP. (DEG. F.)	NUMBER OF AREA ELEMENTS	AREA (SQ. FT.)
1	56.99–58.00	1237	221250
2	58.00–60.00	1521	272050
3	60.00–62.00	1836	328400
4	62.00–64.00	2129	380810
5	64.00–66.00	1555	278130
6	66.00–68.01	1088	194610
7	68.01–69.99	742	132721
8	69.99–72.00	622	111256
9	72.00–74.00	536	95873
10	74.00–76.00	131	23431
11	76.00–78.00	169	36228
12	78.00–80.00	0	0
13	80.00–82.00	0	0

AREA OF ONE ELEMENT (SQ. FT.) = 178.87
TIME OF PASS = 1620
ALTITUDE (FEET) 700
HEADING (DEG. FROM NORTH) 170
SPEED (MPH) 100
GRID SIZE IN FEET = 250

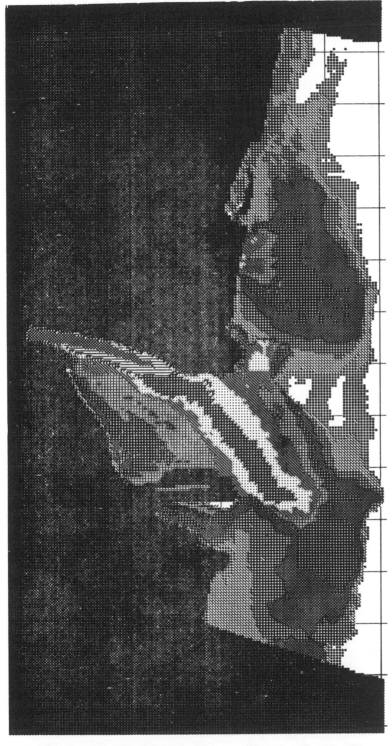

Fig. 11.12

LEVELS 1 2 3 4 5 6 7 8 9 10 11 12 13

or thermal plume temperature, the voltages stored on magnetic tape can be converted by computer to detailed surface concentration or temperature patterns as indicated in figures 11.6 and 11.12, respectively.

A detailed description of the remote sensing instrument systems can be found in the reference cited as note 4 to this chapter.

Reference notes

1. D. J. Baumgartner and D. S. Trent, *Ocean Outfall Design—Part I, Literature Review and Theoretical Treatment*, United States Department of the Interior, Federal Water Quality Administration (April 1970).
2. C. A. Oster, J. C. Sonnichsen, and R. T. Jaske, "Numerical Solution to the Convective Diffusion Equation," *Water Resources Research* 6, no. 6 (December 1970): 1746–52.
3. Alex N. Diachishin, "Dye Dispersion Studies," *Journal of the Sanitary Engineering Division* (Proceedings of the ASCE, January 1963): 29–49.
4. J. R. Eliason, H. P. Foote, and M. J. Doyle, "Remote Sensing Techniques for Tracing the Movement of Industrial Wastes in Surface Waters" (Paper for publication in the *Proceedings of the Pacific Northwest Industrial Waste Management Conference, 1970*).

12. Present Reactor Systems and Their Emissions

J. G. Terrill, Jr. and W. D. Fletcher
Westinghouse Electric Corporation

We will present a quantitative assessment of the exposure of the general public to radiation resulting from the operation of nearby nuclear power plants. In order to determine the existence or magnitude of public health problems which might result from locating many nuclear power plants in a given area, it is important to determine the average exposure levels which might result to the general population from the plants. Throughout this chapter, the information used will be that obtained from Pressurized Water Reactors (PWRs) designed by Westinghouse Electric Corporation.[1] The information on Boiling Water Reactors (BWRs) will be that developed by the Public Health Service in their studies of the Dresden Plant.

In the past, the problem of public exposure to radiation from nuclear plants has been studied for two cases: (1) exposure of the general population following hypothetical accidents, and (2) exposure at the site boundary during normal plant operation.

The first case must, of course, still be studied. The second case, however, must now be expanded to determine the exposure to the *general* population, not just to a *single* individual at the site boundary. The general population may be located at varying distances from the plant sites.

In order to determine the *average* exposure to the general population from nuclear plants, calculations must be analyzed in a different manner, that is, a set of most probable conditions must be used.

These most probable conditions will be applied to the following three sources of radiation exposure to the general public from the use of nuclear

217

power: (1) exposure to the general public from normal operation of a single nuclear plant; (2) exposure to the general public from normal operation of multiple plants; and (3) long-term exposure to world populations from the total nuclear cycle.

From the calculations presented, we will draw conclusions on the amount of average exposure received by the general population from the above radiation sources.

RADIATION EXPOSURE FROM A PRESSURIZED WATER PLANT

The radioactive releases for a typical (1,000,000 kilowatt) PWR of present design are used in the calculations. The results (total body radiation) are presented in table 12.1. On the design basis, the design exposure at the fence line is roughly a factor of almost 100 higher than would actually be anticipated. In the low population zone, five miles downwind and downstream from the plant, the ratio between Federal Radiation Council (FRC) Guide on exposure limits and design basis exposure is a factor of more than 1000. The actual anticipated exposure in this region is roughly a factor of fifty below the design basis. The same general comparisons can be made with the general population zone twenty miles away where the anticipated

TABLE 12.1
Annual Dose from Single Plant
(mrem/year)

		Site Boundary	Low Population Zone (5 miles)	General Population Zone (20 miles)
FRC guide	Air & water *	500	170	170
	background	130	130	130
		630	300	300
Design basis	Air	5.0000	0.1040	0.0156
	water *	0.2055	0.0103	0.0093
	background	130.	130.	130.
		135.2055	130.1143	130.0249
Actual expected	Air	0.0063	0.0001	0.0000
	water *	0.0435	0.0022	0.0020
	background	130.	130.	130.
		130.0498	130.0023	130.0020

* Includes food chain.

exposure is about one-one hundred thousandths of the FRC guidelines. The total whole body exposure to the general public from a single plant is anticipated to be no greater than 0.05 mrem/yr. and more probably in the range of 0.002 mrem/yr. The analysis of individual body organs indicates that their exposure is at least an order of magnitude below total body exposure.

In order to interpret these results and assess their sensitivity to change, it is necessary to identify the assumptions and calculational methods used in their development.

The pressurized water reactor plant releases low levels of radioactivity to the air and to condenser water. The liquid discharges are first processed through a waste disposal system which provides for the removal of a large percentage of each radioactive isotope except tritium. (One of the conservative assumptions used in the design of radioactive waste handling systems for nuclear power plants is that the reactor fuel has defects that release 1 percent of the fission products. This is referred to as the design basis.) The gaseous discharge design basis also includes the assumption of 1 percent fuel defects and a forty-five-day hold up of all gaseous activity to eliminate short-lived radioactive isotopes.

The data on the actually expected exposures were developed by scaling data from presently operating PWRs of Westinghouse design. The anticipated release of gaseous fission products was obtained by comparing actual operation with design basis at several operating PWRs. These experiences have been fairly uniform and show that the 1 percent fuel failure basis gives an overestimate of gaseous releases by a factor of 800 or more. Therefore, gaseous releases for the expected exposure calculations were reduced by this factor of 800 from design basis releases.

The exposure from airborne activity, an external source of radiation, assumes that a person is submersed in a cloud of the radioactive gases. The local concentrations of radioactivity were obtained by procedures outlined in Pasquill.[2] A ground level discharge was assumed with stable, neutral, and unstable atmospheric conditions occurring 40 percent, 30 percent and 30 percent of the time, respectively, and 4.5, 13.4, and 6.7 miles per hour average wind speeds, respectively.

The liquid discharge exposure calculations include tritium releases and nontritium releases, both of which are internal sources of exposure. The tritium design basis calculations contain the assumption that 30 percent of the tritium from the fission process is released to the coolant and that 100 percent of the tritium from control rods and boric acid is produced and/or released in the coolant. The anticipated tritium release level includes the assumption that there will be substantially less tritium released with fuel clad with zirconium than stainless steel.

If this anticipated reduction does not occur, water exposures will be somewhat higher than those shown as "anticipated" and, as an upper limit, the anticipated water exposure levels will approach design basis levels.

Nontritium liquid releases, presently termed unidentified isotopes, have not been as consistent among operating PWRs. For the expected release analysis, it was assumed that the unidentified isotopes were six times that given in the design basis. This factor of six was applied uniformly to all liquid isotopes except tritium. There may be some evidence that any excess activity in the water above the 1 percent design basis is not from fission products but may be principally due to activated crud (normal oxidation of surface metals). In this case, the anticipated exposures from water and food chain concentrations will be slightly reduced.

The exposure from liquid releases has two components: exposure from direct intake of drinking water from the river and exposure from solids and water intake through aquatic foods. The aquatic foods could include fin fish, mollusks, and crustacea associated with commercial saltwater fishing, but in the freshwater system, fin fish are the only fish food available. The calculations, however, did not make this distinction. The assumed diet composition was obtained from a weighted population average and the total United States aquatic food catch in 1966. The contribution of exposure from aquatic foods was substantially less than 5 percent of the water intake dose.

RADIATION EXPOSURE FROM MULTIPLE PLANTS

The assumptions and calculational procedures are the same for the multiplant model as for the previous case except for the zones of overlapping radiation.

A simplified model can be developed involving three 1,000,000 kilowatt PWR-type plants having overlapping exposure zones within a twenty-mile radial distance of each other. Exposures are calculated at the four zones: (1) downstream from three plants and distances of five, thirty, and twenty-five miles from each of the three plants, respectively; (2) downstream five miles from one plant and about twenty miles upstream from two other plants; (3) downstream from all three plants, essentially twenty miles or more from the nearest reactor plant; and (4) upstream but within the air shed of all three plants, at distances of eight, twenty-eight, and thirty miles.

The calculated exposures are shown for each point on the basis of design and anticipated values in table 12.2. In this table, it can be noted that the incremental radiation exposures through all mechanisms provide, at most, an average exposure to the general public of 0.003 mrem/yr. and as little as 0.001 mrem/yr. or less. This is extremely low compared to the allowable guideline exposure from nonmedical man-made radiation of 170 mrem

TABLE 12.2
Annual Dose from Multiple Plant Sites

		Zone 1	Zone 2	Zone 3	Zone 4
FRC guides	Air & water *	170	170	170	170
	background	130	130	130	130
		300	300	300	300
Design basis	Air	0.1040	0.1040	0.0156	0.1040
	water *	0.0103	0.0144	0.0139	0.0000
	background	130.	130.	130.	130.
		130.1143	130.1184	130.0295	130.1040
Actual expected	Air	0.0001	0.0001	0.0000	0.0001
	water *	0.0022	0.0031	0.0030	0.0000
	background	130.	130.	130.	130.
		130.0023	130.0032	130.0030	130.0001

* Includes food chain.

above background established by the Federal Radiation Council.[3] In any other field of investigation, these anticipated exposure levels would be considered minor perturbations from natural background and thereby equated to zero.

RADIATION EXPOSURE FROM A BOILING WATER REACTOR

A detailed study by the Public Health Service on the Dresden I Plant of the Commonwealth Edison Company provides the most complete data on a boiling water plant.[4] Dresden I is a boiling water reactor (BWR) that has generated more than 9×10^9 kilowatt-hours since it began operation in 1959, and has been operating at a rated power of 700 thermal megawatts (Mwt.) and 210 megawatts of electricity (Mwe.) since 1962. The power plant is a dual-cycle, forced circulation system, built by the General Electric Company. Fuel elements consist of slightly enriched UO_2 clad in Zircaloy-2. The station is located in Illinois, 80 km. southwest of Chicago. Liquid wastes are discharged into the cooling water discharge canal which empties into the Illinois River; gases and airborne particles are released from a 91-m stack; and solid wastes are transported off-site for burial.

The study was planned and performed by the staff of the Division of Environmental Radiation, Bureau of Radiological Health, Environmental Health Service, PHS, with the close cooperation of the Illinois Department of Public Health, the Commonwealth Edison Company, and the Division of Radiation Protection Standards of the AEC. The data were collected in five field trips on 14–16 November 1967; 25–28 January 1968; 31 January–1 February 1968; 25–29 June 1968; and 20–22 August 1968.

Emphasis was placed on relating discharges to environmental levels of radionuclides and radiation, and on evaluating critical pathways and radionuclides. The greatest efforts were devoted to (1) measuring the concentration of individual radionuclides in liquids and gases on site and in effluent liquids, gases, and airborne particles; (2) determining the radiation exposure rate and radionuclide concentration in air at ground level beneath the plume from the stack; and (3) testing devices that concentrate radionuclides in air and water for subsequent analysis.

GASEOUS FISSION PRODUCTS

Average release rates of the measured noble gas fission products are listed in table 12.3.

TABLE 12.3
**Average Release Rates of Measured Noble Gas
Fission Products**

Fission Product Isotope	Effluent (μCi/sec.)
4.4-hr. ^{85m}Kr	3×10^2
10.7-yr. ^{85}Kr	1×10^{-1}
76 -min. ^{87}Kr	7×10^2
2.8-hr. ^{88}Kr	5×10^2
2.3-d ^{133m}Xe	1×10^1
5.3-d ^{133}Xe	3×10^2
9.1-hr. ^{135}Xe	8×10^2
17 -min. ^{138}Xe	2×10^3

The average values were computed for the 1968 average fission-product noble gas release rate of 12,500 μCi/sec. during 64 percent of the year. These values were generally in accord with measurements performed earlier by General Electric Company staff.

Table 12.4 shows average release rates of the other measured radionuclides.

The measured release rates were consistent with the average annual releases reported by Dresden Nuclear Power Station for the year 1968 of 0.189×10^{-7} μCi/ml of water and $12,500 \times 0.64 = 8,000$ μCi/sec. in air. The sum of measured noble gas fission products (see table 12.3) is 4,600 μCi/sec.; the gases ^{83m}Kr, ^{89}Kr, ^{135m}Xe, ^{137}Xe, and ^{13}N would be expected to contribute additional activity amounting to several thousand μCi/sec.

The influence of plant operation on radionuclide release rates in effluent is indicated by comparing these release rates to fission production rates in the reactor core. The ratios of the fission product release rates (table 12.3) to estimated production rates are found in table 12.5.

TABLE 12.4
Average Release Rates of Other Measured Radionuclides

Radionuclides	In Stack Effluent (μCi/sec.)	In Liquid Effluent (μCi/sec.)
^{3}H	6×10^{-3}	5×10^{-2}
^{58}Co	2×10^{-5}	2×10^{-2}
^{60}Co	2×10^{-5}	3×10^{-2}
^{89}Sr	7×10^{-4}	8×10^{-3}
^{90}Sr	3×10^{-6}	9×10^{-4}
^{131}I	6×10^{-4}	1×10^{-3}
^{134}Cs	$<1 \times 10^{-5}$	2×10^{-3}
^{137}Cs	2×10^{-5}	6×10^{-3}
^{140}Ba	3×10^{-4}	5×10^{-3}
^{144}Ce	$<3 \times 10^{-5}$	2×10^{-4}

TABLE 12.5
Ratio of Fission Product Release Rates To
Estimated Production Rates

Noble Gas Fission Products		Other Fission Products	
^{85m}Kr	1×10^{-6}	^{3}H	1×10^{-3}
^{85}Kr	5×10^{-5}	^{89}Sr	4×10^{-9}
^{87}Kr	6×10^{-7}	^{90}Sr	6×10^{-8}
^{88}Kr	7×10^{-5}	^{131}I	2×10^{-10}
^{133m}Xe	6×10^{-5}	^{137}Cs	4×10^{-7}
^{133}Xe	1×10^{-5}	^{140}Ba	4×10^{-10}
^{135}Xe	2×10^{-6}	^{144}Ce	4×10^{-10}

Thus, only small fractions of these fission products are released. The ratio for fission-produced tritium is actually lower than shown because most of the discharged tritium is probably produced by neutron activation of deuterium in reactor cooling water.

EXPOSURE FROM RADIONUCLIDES AND RADIATION IN THE ENVIRONMENT

The main effort in the environmental aspect of this study was devoted to translating radionuclides in the plume from the Dresden stack into the external radiation exposure from these radionuclides. The plume was detected as much as 18 km. distant from the stack with large NaI (TI) survey instruments. Within 1 to 2 km. from the stack, radiation exposure rates were measured at the centerline of the plume under stable and neutral

conditions with a tissue-equivalent ionization chamber and sensitive elec-
trometer. Average exposure rates measured during three tests for one-half-
to one-hour periods were 13, 24, and 40 μR/hr. Concentrations of ^{133}Xe,
^{135}Xe, and ^{138}Cs in ground-level air were measured at the same time as the
radiation exposure rates. Estimated concentrations and exposure rates,
based on measured release rates at the stack and diffusion calculations, were
within a factor of two or better of the measured values.

Thermoluminescent dosimeters exposed during two-week periods at ten
stations, located between 1 and 4 km. from Dresden, showed an average
background radiation exposure rate of 9 μR/hr. During reactor operation,
TLDs at three locations indicated radiation exposure from the plume at
average rates for a two-week period of 2 to 3 μR/hr. above the natural
radiation background. The TLD values are uncertain, however, because
their standard deviation is $\pm$ 1 μR/hr., and variation in the background
could have introduced significant error.

Samples collected in the Dresden environment for analysis of individual
radionuclides usually contained radionuclides such as ^{90}Sr and ^{137}Cs from
fallout associated with atmospheric testing of nuclear devices, and naturally
occurring radionuclides such as ^{40}K, ^{232}Th plus progeny, and ^{226}Ra plus
progeny. In order to search for radionuclides attributable to Dresden,
nearby samples were compared with more distant samples, and downstream
or downwind samples with those collected upstream or upwind. In addition,
measurement of ^{58}Co, ^{60}Co, ^{89}Sr, ^{131}I, and ^{134}Cs was emphasized, because
these were discharged at Dresden while their concentrations in fallout were
low or undetectable.

No radioactivity attributable to Dresden was found in the samples of
rainwater, soil, cabbage, grass, corn husks, milk, deer, rabbit, surface water,
drinking water, or fish. Estimated concentrations and measured less-than
values suggest that, for example, a sensitivity of approximately 0.05
pCi/liter appears necessary to measure ^{131}I from Dresden in milk and 0.001
pCi/liter to measure ^{89}Sr from Dresden in rainwater.

Indications of radioactivity from Dresden were found in the following
samples:

1. The average ^{131}I content of three thyroids of heifers was 31 pCi (0.45
   pCi/g). These heifers had been placed on pasture 2.3 km. east of
   Dresden for several weeks before slaughter specifically for testing
   cattle thyroid as a sensitive indicator of ^{131}I deposition on pasture
   grass.
2. Snow collected 0.9 km. south of Dresden contained ^{89}Sr at a concentration of 10 pCi/liter, while snow collected 10 km. north-northeast
   of Dresden contained no ^{89}Sr ($<$1 pCi/liter). The area to the south
   of Dresden was downwind during the snowfall.

3. Kernels of field corn grown at the same location south of Dresden contained ^{137}Cs at a concentration of 4 pCi/g ash (at 0.8 percent ash weight), compared to 0.2 to 0.8 pCi/g ash in corn kernels from other locations.

On the basis of these measurements, exposure to the surrounding population through consumption of food and water from radionuclides released at Dresden was not measurable. External exposure from radioactive gases discharged from the Dresden stack was detectable, but it was only a small fraction of the natural radiation background over an extended period of time and well within Federal Radiation Council guidance.

LONG-TERM BUILDUP IN THE ENVIRONMENT

In addition to the emissions of some long-lived radioisotopes at the reactor, recognition must be given to the question of long-term buildup in the environment. Since the isotopes involved are created in the same proportion to the energy produced, these remarks apply equally to the pressurized and boiling water reactors.

When the nuclear fuel has completed its function at the nuclear power plant, it is shipped to a chemical reprocessing plant where the fuel is chemically dissolved and separated into fuel material and fission products. The fuel reprocessing plant provides extensive treatment of the gaseous and liquid effluents, removing large fractions of all radioactive isotopes except tritium and the noble gases, krypton, and xenon. Currently, these isotopes (representing 3000 parts per million of all radioactive isotopes produced) are released to the environment during reprocessing. This is approximately 1000 times the quantity released at the reactor plant.

Two of these isotopes, krypton and tritium, have half lives of about ten years; there is, however, no known or conceivable mechanism for reconcentration of these radioactive isotopes in nature above release levels. It would appear that these radioactive isotopes will tend, in time, to be distributed more or less uniformly throughout the biosphere. While the planetary buildup of these radioactive materials does not pose a near-term problem, it is important to calculate the long-term situation in order to determine if, when, and what kind of additional waste retention systems may be required.

The long-term study of the buildup of these isotopes has been the subject of previous studies reported in the literature which we will summarize. The assumptions used in these long-term models involve a continuing increase in electrical power requirements for the world, with nuclear power accounting for one-half of total electrical generation capacity by the year 2000 and increasing still further in years beyond. It is also assumed for the projection of tritium releases that all future reactors are of present PWR design and all future reprocessing plants operate as at present.

The resulting buildup of reactor-produced tritium increases until it has reached the level of cosmic ray–produced environmental tritium in the late 1980s and will, around the turn of the century, be equal to the residual tritium from the hydrogen weapons tests of the last two decades. After the year 2000, reactor-produced tritium could become the dominant source of tritium in the environment if it is not retained and contained by that time. In the year 2000, the exposure to the world population from all sources of tritium would be 0.002 mrem.[5]

For the long-range environmental buildup of krypton, a changing mixture of reactor types, including thermal converters and fast breeders, is assumed. Estimates of the worldwide exposure, which would result from the long-term buildup of krypton, indicate that average exposures to the world population would be between 50 and 100 mrem/yr. one hundred years from now.[6] While this exposure level is within internationally recommended guidelines, it would seem prudent for the worldwide nuclear power interests to undertake the necessary corrective measures, particularly at reprocessing plants, in order to minimize the impact of nuclear energy on the total environment. There would appear to be ample time to complete the development of krypton retention systems and proceed with incorporating them into all reprocessing plants constructed after the present decade.

ENVIRONMENTAL IMPACT

As the record clearly indicates that nuclear power plants have fully met the regulatory requirements, why should there be any public concern about the environmental impact of nuclear power? In the authors' opinion, this concern can be divided into two categories: one is related to science and technology and the other is related to public expressions and attitudes. Let us consider examples of the types of legitimate technical and scientific concern which should be under continuing review. The first item is the basic guidelines and recommendations of the Federal Radiation Council and the National Council on Radiation Protection and Measurement and its international counterpart. Assuming as both organizations do, that there is no threshold for some types of radiation effects, is it proper for groups largely composed of scientists to make these value judgments? The Federal Radiation Council mechanism was created to assure input of the social as well as the physical sciences; but in the minds of many, it has not effectively achieved this goal. Presently, with the consolidation of the Federal Radiation Council and the Environmental Protection Agency, this method of publicly assuring the input of the social sciences will no longer be readily feasible.

Another type of concern is related to the accumulation of data which indicate that certain types of materials which have been assumed to be perfectly safe, such as DDT, are found to be creating environmental problems which were not clearly recognized by the scientific experts for many years. Many conscientious people can be justifiably concerned about the capabilities of the experts in predicting future radiation effects. In this instance, however, radiation effects have been studied much more thoroughly than pesticides and the radiation background has been known for thousands of years, whereas the same level of research and natural exposure to DDT has not been available for comparative purposes.

Another legitimate concern is the effect of the total nuclear generating system including mining, fuel reprocessing, and ultimate disposal of high-level and long-lived wastes. The mining situation is rapidly being improved. Environmental protection improvements are also being made at fuel processing plants. We have many years in which to further develop our capabilities of handling the radioactive wastes with long half-lives. In each instance, however, the nuclear industry must recognize that a continuing effort directed at making the entire industry a model industry from the standpoint of environment, as well as economics, will be required to answer the legitimate public questions about nuclear technology.

Another group of critics utilize the apparent inability of the industry to provide public information in a form that is consistent with scientific fact. Spokesmen for the industry make such statements as, "The nuclear industry has never compromised with safety." However, the basic Federal Radiation Council guidelines are by definition a compromise. When statements of this type appear in the public media, they provide critics with an opportunity to claim that the industry does not recognize the basis upon which their standards are derived. In the heat of publicity generated by the incredibility of industry statements, the overall conservatism of the standards and operational records of the nuclear industry will appear only in the fine print of public releases, if at all.

The nuclear industry has perhaps the best environmental record of any major industrial development to date, judging from health and scientific studies. However, some long-range problems, such as long-term effects of radiation and ultimate disposal of long-lived radioisotopes, require our continuing study and development.

Reference notes

1. J. H. Wright, "Environmental Radiation from Pressurized Water Reactors," *Atomic Power Digest* (Pittsburgh, Pa.: Westinghouse Nuclear Energy Systems, 1st Quarter 1970).
2. F. Pasquill, "Atmospheric Diffusion" (London: D. Van Nostrand Company Ltd., 1962).

3. Federal Radiation Council, "Background Material for the Development of Radiation Protection Standards," Report no. 1 (Reprint, Department of Health, Education, and Welfare,
   13 May 1960).
4. B. Kahn, et al., "Radiological Surveillance Studies at a Boiling Water Nuclear Power
   Reactor," BRH/DER 70-1 (Public Health Service, Department of Health, Education, and
   Welfare, March 1970).
5. H. L. Price, "Background Information on Release of Radioactivity in Nuclear Power
   Reactors," in *Selected Materials on Environmental Effects of Producing Electric Power,*
   Joint Committee on Atomic Energy, Congress of the United States (United States Government Printing Office, August 1969).
6. J. R. Coleman and R. Liberace, *Radiological Health Data and Reports* 7, no. 11 (United
   States Department of Health, Education, and Welfare, Public Health Service, November
   1966).

13. Future Reactor Systems

Harry Lawroski
Argonne National Laboratory

Extensive concern about the availability and quality of energy for the development of the world exists today. The exponentially increasing world population and its tremendous appetite for energy will present problems of differing degree, depending primarily upon the location and desired standard of living. Certainly, as time passes, the various peoples of the world will attain a standard of living similar in many respects. The methods and rates of progress will, however, vary widely.

On a national level, standard of living is closely attuned to per capita energy consumption at present. With a relatively advanced society the types of energy presently used vary from the gasolines or oils for vehicles to the coals and oils, the hydro potentials, and the nuclear fuels for stationary power plants. In more underdeveloped countries, the fuel for cooking may be the principal energy expended. As the more developed nations attain greater social conscience, they will help backward nations reach their standard of living.

It is not conclusively known how much energy is truly needed. How much are individuals willing to commit to creature comforts? Each thinks differently. It is certainly not practical to take a vote to decide; knowledgeable, conscience-bound people must proceed to formulate plans and implement progress.

ENERGY RESOURCES

An examination of the literature, which is copious, can only offer guesstimates at best. History has proven that there are usually more resources

229

available than the initial estimates in reserves of natural resources indicated; however, it would be unwise to rely on this. I shall give my own consensus of the information in this chapter.

The gross energy consumption of the world in 1960 was 0.15 Q, 1970 was 0.25 Q, 1980 is expected to be 0.6 Q, 2000 is expected to be 1.3 Q, and 2020 is expected to be 2.5 Q, where $Q = 10^{18}$ Btu. At present it appears the United States is using approximately 40–50 percent of the total. This portion must change and eventually even out as the rest of the world improves its standard of living.

Where are we going to get our energy? Even if fossil-fuel reserves exist, should we utilize them for the purpose of making power or for other, better purposes? World energy reserves are: fossil 450 Q, fission 5,000,000 Q, and fusion 10,000,000 Q, where $Q = 10^{18}$ Btu. Translating this data into years, there exist approximately ten to fifteen years of energy from presently known economic fossil fuels, and about 200 years assuming utilization of *all* of the fossil fuel including that not yet discovered.

As in the case of fossil fuels, the cost is important in estimating near-term and future sources of nuclear energy. For a cost of $5–10 per pound of U_3O_8, the known reserve is about twenty-five years. Using the fast breeder with the $5–10/pound U_3O_8, the known reserve will last approximately 300 years. With new discoveries and more expensive ore, an extrapolation of 10^6 years is possible. Considering fusion, the time period might be approximately 2×10^6 with the lithium-6 supply and 2×10^9 using the deuterium fusion system. Judging from these considerations, the necessary energy resources exist; however the manner in which we use them is up to us.

There are various ramifications to the increased demand for power: (1) an increase in waste heat; (2) an increase in the cost of land for plants and for transmission of power; and (3) an increase in transportation of fuels and refuse.

Since power plants are getting larger, the unusable heat energy localizes in the area of the plants. The local environment can accept the perturbation, or more efficient plants can be built, or the waste heat can be used for other purposes such as warm water agriculture. In the latter two cases, the initial cost of the system increases significantly.

THERMAL REACTORS

At present, the water-cooled reactors utilizing enriched uranium are competitive with other sources of energy for power plants. For the next ten to fifteen years the water reactors will be the most efficient for adding new electrical generating capacity.

Let us briefly compare various reactor systems. Figure 13.1 is a simplified

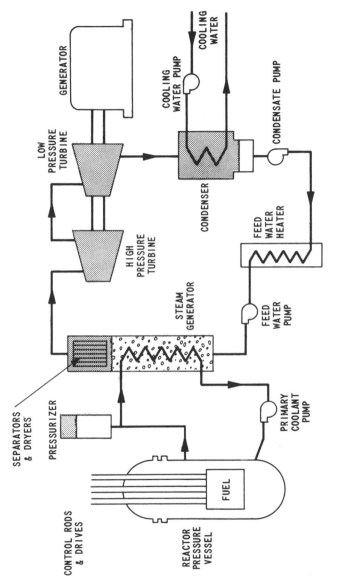

FUEL Slightly enriched uranium oxide clad with zirconium alloy

MODERATOR Water

COOLANT Water

PRESSURE OF PRIMARY SYSTEM 2,250 pounds per square inch

REACTOR OUTLET TEMPERATURE 605°F

Fig. 13.1 Pressurized water reactor power plant.

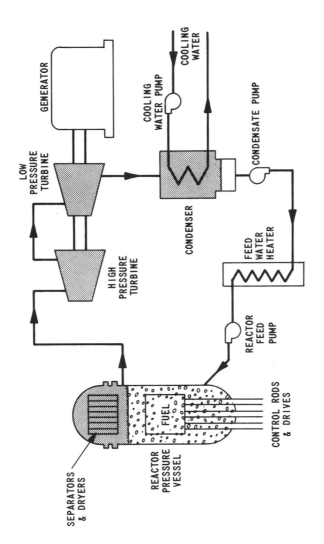

FUEL Slightly enriched uranium oxide clad with zirconium alloy
MODERATOR Boiling water
COOLANT Boiling water
PRESSURE OF PRIMARY SYSTEM 1,000 psi
OUTLET TEMPERATURE 550°F

Fig. 13.2 Boiling water reactor power plant.

schematic of a pressurized water reactor power plant. It is basically a double-circulating system. Although the primary coolant system operates at 2250 to 2500 psi and the exit coolant temperature is in excess of 600° F., there is a degradation of energy level because of the heat exchange to the secondary system, resulting in steam at 550 and 575° F. and 1000–1250 psi delivered to the turbine.

A comparable boiling water reactor power plant is depicted in figure 13.2. This shows a single-cycle system in which the steam pressure is at 1000 psi with an outlet temperature of 550° F.

Since neither system operates at high temperatures, the efficiency suffers. Most water reactors operate at about 30–33 percent overall efficiencies.

Another thermal reactor concept which has been slower in developing is the high temperature, gas-cooled reactor. Within the next year, a 330 Mwe. unit will be operating. This type of reactor produces high quality superheated steam and should have a thermal efficiency of nearly 40 percent. The higher efficiency would therefore present a relatively lower rejection of waste heat to the environment.

FAST BREEDER REACTORS

One of the criteria for deciding whether to pursue a program is the benefit/cost ratio. Recently the United States Atomic Energy Commission evaluated the fast breeder program with respect to building a demonstration plant and then proceeding to larger plants. The calculated benefit/cost ratio was 9/1, which would indicate the desirability of encouraging this program.

The major advantages of the fast breeders over the water reactors are (1) higher energy extraction efficiency and (2) better utilization of resources. A reasonable rule of thumb is to figure that breeder reactors should extend the usable uranium reserve by a factor of 100.

The basic flow plan of a fast reactor is shown in figure 13.3. The system is a three cycle system: (1) primary sodium, (2) secondary sodium, and (3) steam. The temperatures exiting from the reactor are some 400 to 500° F. higher than those in water reactors so that, even with the extra heat-exchanger degradations, the resulting steam temperatures are 400° F. higher than those in water reactors. These higher temperatures present a higher efficiency of approximately 40–42 percent.

Presently the Experimental Breeder Reactor at the Argonne National Laboratory operates at an outlet sodium temperature of 883° F. with the inlet temperature of 700° F., producing steam above 800° F. at 1250 psi to the turbine. The initial LMFBR demonstration plants will have outlet temperatures in the range of 1000–1050° F., which should result in overall thermal efficiencies of slightly under 40 percent.

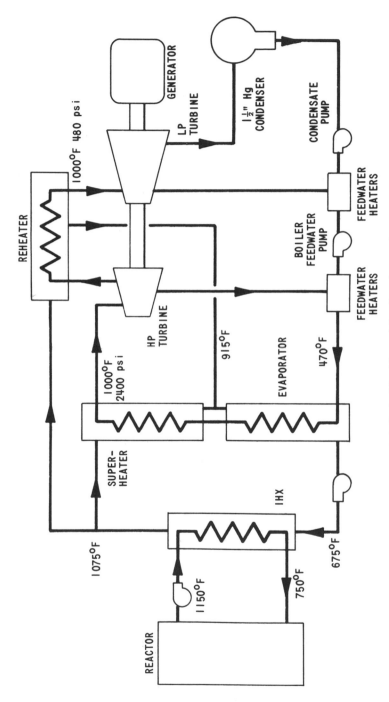

Fig. 13.3 Typical liquid metal fast breeder reactor.

234

There are two basic coolant-system concepts being pursued in the LMFBR program: (1) pool and (2) loop. Since the power density and fuel concentration in the fast breeder using sodium coolant are considerably higher than with water, the decay heat during reactor shutdown is substantial. For this reason shutdown cooling is more stringent and must be continued for some time after shutdown.

The pool type provides a substantial heat capacity for cooling, but requires the primary pumps and the intermediate heat exchanger to be immersed in the primary tank sodium. This in turn makes the maintenance of these components more complicated. Experience at EBR-II shows that with proper planning and preparation the pumps can be removed, cleaned, inspected, and returned to service. The reactor vessel is larger to accommodate the various components.

The loop type requires that specific consideration be given to shutdown cooling, in that auxiliary cooling loops or extra motors are provided. It is of interest to note that access to the primary pumps and the intermediate heat exchanger is normally better in this type of cooling system.

At present the British, the French, and the Russians are building pool types of demonstration fast breeder reactors. The United States reactor suppliers are considering both types for the demonstration plants. At this time it is certainly not clear which type is superior; both have advantages and disadvantages.

The fast breeder reactors will aid the environment by reducing the proportion of exhaust heat. The emission from an LMFBR will be less than the present emissions from the water reactors.

The LMFBRs will contain substantial amounts of plutonium, for example, a 1000-Mwe. plant is estimated to require approximately 2500–2800 kg. of plutonium. The containment aspects of these quantities at these plants should be no greater than the present problems of containment at any reactor. Good engineering and safe practices should prevent releases.

The shipping of expended fuel to distant processing plants is of more concern. The first few demonstration plants will probably handle shipping by partially dismantling subassemblies at the reactor site. However, as more reactors are built, the transportation of plutonium in this country may be substantial. A scoping estimate is that 3×10^6 Mw. may be needed in the United States in the year 2000. Assuming half of this is in fast reactors and if they are all 1000 Mwe., this indicates perhaps 1200 reactors. With these numbers there will be about 3×10^6 kg. of plutonium being shipped yearly. The logistics become monumental. It appears that a better solution would be to start planning and providing for power parks which will have plants for the processing and remanufacturing of the fuel near the power stations.

Obviously the power parks will either have to be near water for cooling or they must provide air heat exchangers. It staggers the imagination to envision the cooling towers required in either case. Further, these power parks could change the ecology downwind either for the better or for the worse. A power-park study should proceed in the near future. Further, it is necessary that an overall look at our future power scheme be initiated.

FUSION

The promise of fusion must not be ignored. Its proponents, however, should try to be practical in the approach. Basically, the fusion reactor is many years in the future but there have been several important advances. The Tokamak and Scyllac machines are good examples.

The basic equations for the fusion reactors are:

$$_1D^2 + {_1}T^3 \longrightarrow {_2}He^4 \ (3.5) + n \ (14.1)$$

$$_1D^2 + {_1}D^2 \to \begin{cases} \to {_2}He^3 \ (0.82) + n \ (2.45) \\ \\ \to {_1}T^3 \ (1.0) + {_1}H^1 \ (3.02) \end{cases}$$

$$_1D^2 + {_2}He^3 \longrightarrow {_2}He^4 \ (3.6) + {_1}H^1 \ (14.7)$$

An examination of the energy inputs indicate the best is the deuterium-tritium reaction. The primary energy comes from the kinetic energy of the 14.1-MeV neutron. Although the products of the reaction are not radioactive, the tritium reactant is; also the activation by neutron collisions will be very substantial.

At this particular time it is not necessary to go into the details of the systems such as the heat loss, the essential energy inputs to produce the reaction, etc. It would be of interest to look at a conceptual schematic plant, which is shown in figure 13.4. Superconducting coils would be used to confine the plasma. The materials of construction are one of the major problems even if the sustained reaction is achieved. Neutron damage to the reactor components would be substantial and frequent replacements probably would be necessary. These structural materials in all probability would be highly activated and difficult to handle.

The coolant and neutron absorber is shown to be a lithium salt. A lithium beryllium coolant has been the subject of more recent consideration. The lithium is needed to obtain the necessary tritium. In addition, very good heat transfer is essential. The sodium technology developed for the LMFBR would indeed be an asset to the future development of the fusion reactors.

The research problems are many. The tritium recovery systems will require extensive research and development. Inventory is substantial. The technology of confinement of large amounts of tritium and the reprocessing method must be worked out before real progress can be made. Another perplexing area is the injection of the plasma into the reactor.

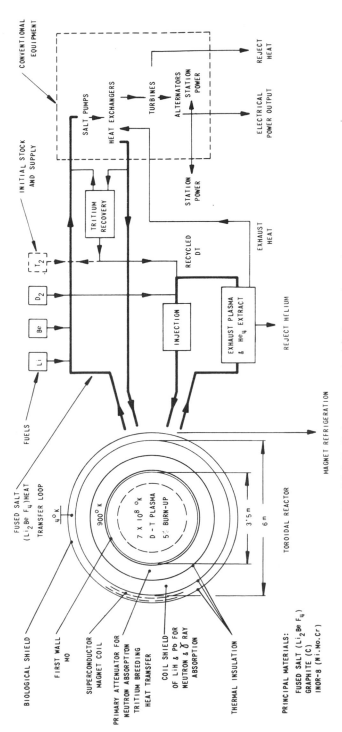

Fig. 13.4 Preliminary conceptual fusion reactor system.

237

The second two equations present an interesting possibility: essentially no waste exhaust heat, better known as direct conversion. Theoretically, the protons given off could be converted directly into electricity. Unfortunately the condition to obtain the reaction profitably requires considerably more stringent plasma conditions and will be harder to achieve.

The fusion reactor will require the equivalent of criticality for the fission reactor. This is probably one to two decades in the future. After the sustained reaction, a feasibility reactor will be needed, and then a larger demonstration unit. Fortunately, much of the coolant technology is now being developed as part of the LMFBR and could be applied to the fusion heat removal systems.

SOLUTION TO ENERGY NEEDS

The energy requirements of the world continue to grow at an increasing rate. Present conventional power systems are taxing the energy reserves; future advanced power systems are needed. These systems should be designed to extend the usable energy sources, while minimizing the insult to the environment. Consideration must be given to the efficiency of the systems, the distribution of power, the effluents from the operating systems, the fuel processing, and the management of the waste products.

The technology for fast breeder reactor systems is essentially at hand, and operating plants have demonstrated its feasibility. Development to increase the sizes to permit economic utilization of the fast breeders which will extend the energy resources of the world, is proceeding at a studied rate. The handling of large amounts of plutonium and radioactive wastes must be accomplished in a safe, economical manner to minimize environmental problems. Reasonable projections show that the fast breeder reactor is the near-future and perhaps the far-future solution to the energy needs of the world; it is the best candidate for the next fifty years of power requirements. Certainly a dark horse for the far-future solution is the fusion reactor.

In any event, it is believed that power plants will have to grow in size and the future power stations should be sited in a "power park," which would include the fuel-processing plants. These power parks could quite easily be in remote areas away from population centers.

14. Fuel Processing and the Environment

Albert W. De Agazio
NUS Corporation, Environmental Safeguards Division

A concern frequently expressed by conservationists at reactor licensing proceedings is the potential threat to our environment resulting from the reprocessing of reactor fuels. The reprocessing of spent fuel elements from nuclear power reactors is certainly one aspect of the nuclear fuel cycle which, if not properly controlled, could have a major effect on environmental radioactivity levels. Commercial power reactors, currently in operation or under construction in the United States are fueled by slightly-enriched uranium; that is, fuel in which the percentage of the uranium-235 isotope in the uranium mixture is higher than the 0.7 percent found in nature. When fuel is replaced after perhaps three years exposure in the reactor, there is still a significant amount of U-235 remaining in the spent fuel elements; in addition, some fissionable plutonium isotopes are also created during the exposure of the fuel in the reactor. These so-called spent fuel elements contain, in fact, a substantial value of unburned uranium-235 and plutonium, and are transferred to fuel reprocessing facilities for the recovery of these values from the spent fuel.

The fissioning of a uranium-235 atom produces, in addition to the prompt release of energy, a slower release of energy in the form of radioactive fission products. Although this delayed energy source accounts for some of the useful heat production in a reactor, it also represents a potential hazard to man should these radioactive fission products be allowed to disperse in the environment. In a power reactor, the uranium fuel is in the form of uranium dioxide pellets sealed within a zirconium tube. Thousands of these loaded tubes or fuel elements make up the reactor core. A major concern of both the fuel designer and reactor operator is to minimize the release of fission products from fuel elements since this has a direct effect upon the radiological safety of the facility. During the course of operation of a power reactor, small amounts of the radioactive fission products contained within the fuel elements will escape through defects which cannot be completely eliminated in the canning surrounding the fuel material. Because of this, power reactors are designed to operate with a certain amount of radioactivity circulating with the reactor coolant as a result of operating with a small faction of defective fuel. The amount of radioactivity released from such defective fuel is only a small fraction of the radioactive fission products contained in the fuel elements. Typical reactor systems operating with the maximum design fission product levels in the coolant would only result in about 1/1000th of the fission product radioactivity escaping from the fuel. We should note here that this radioactivity is not released directly to the environment, but is contained within the reactor coolant system. Other systems are installed to remove as much of this radioactivity as possible so that only a small fraction of the radioactivity circulating with the coolant escapes to the environment.

In the fuel recovery process, the fission products which have been so carefully retained within the fuel element during its life in the reactor core are separated from the uranium and plutonium to permit reuse of the latter in new fuel. The control and management of these fission products create significant engineering problems, but problems which have been successfully managed for many years in the Atomic Energy Commission's own fuel reprocessing installations.

The typical power reactor being constructed in the United States today is refueled approximately once a year. When such a reactor is refueled, one-third of the fuel elements—the oldest—are removed and transferred to a storage pool. There the fuel is submerged in water, which provides both shielding and cooling, for a period of from 60 to 150 days. The purpose of the long storage period is to allow most of the shorter-lived fission products to decay. This reduces the heat and radiation levels for which shipping containers must be designed.

After the holding period has expired, the fuel is transferred to the shipping containers. The shipping containers are specially designed units which provide for cooling and shielding the highly radioactive fuel elements during transport. These shipping containers must be capable of withstanding the consequences of any foreseeable accident such as fire or collision without releasing radioactivity to the environment.

At the fuel reprocessing plant, the fuel elements are unloaded from the shipping containers underwater and may be stored for further radioactive decay depending upon how much time has elapsed from removal from the reactor. From here, the fuel is moved to "head end" processing where the fuel elements are prepared for the subsequent processing steps. At the "head end," miscellaneous parts such as end pieces, spacers, and springs are removed. Burnable poison rods which do not contain fuel, if they are used, are also removed.

AQUEOUS SOLVENT EXTRACTION PROCESS

At the present time, there is only one commercial processing plant in operation in the United States. This is the Nuclear Fuels Services (NFS) plant in upper New York state. General Electric's Midwest Fuel Reprocessing plant is under construction near the Dresden site of Commonwealth Edison and a construction permit has been granted to Gulf-Allied's Barnwell South Carolina plant. The NFS plant and Gulf-Allied's Barnwell plant use a modified version of the aqueous solvent extraction process originally developed by the AEC for use in the Commission's separations facilities. The Midwest Fuel Reprocessing plant uses the Aqua-Fluor which differs in the basic separations steps. Let us discuss the aqueous solvent extraction process first. Figure 14.1 is a block diagram of the basic process used at the NFS plant and at the Barnwell plant. After the nonfuel bearing portions of the fuel element have been removed, the individual fuel rods are chopped into small pieces and placed into metal baskets. The basket is then placed into a dissolver where the fuel values, uranium and plutonium, are dissolved together with the fission products in a hot nitric acid solution.

The fuel cladding materials or hulls are monitored to make sure that all fuel is dissolved prior to their disposal as solid radioactive waste. The dissolved fuel and fission products in the nitric acid solution are then contracted with an organic solvent which selectively extracts the uranium and plutonium leaving behind the overwhelming bulk of the fission products in the nitric acid solution. There are several more steps of solvent extraction and other chemical separation processes employed to purify the recovered uranium and plutonium from any fission product contamination remaining and finally to separate the uranium and plutonium from each other; in terms of waste management, however, the most important steps are those at the beginning of the fuel recovery process.

AQUA-FLUOR PROCESS

The Aqua-Fluor process, shown on figure 14.2, also involves first chopping the fuel rods into small pieces, followed by dissolution in hot nitric acid and a solvent extraction step to separate the fuel from fission products.

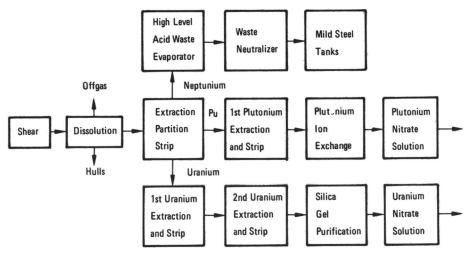

Fig. 14.1 Nuclear fuel services: process description.

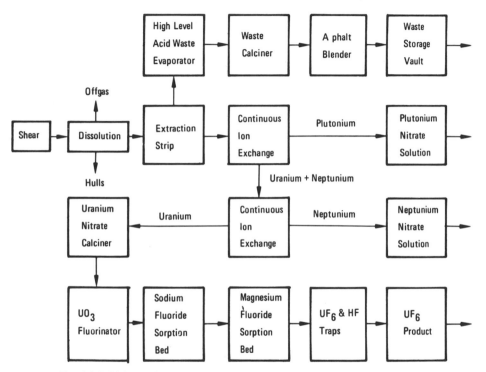

Fig. 14.2 Midwest fuel reprocessing plant: proposed process description.

At this point, the process differs. Subsequent separations are accomplished by ion exchange and sorption beds. However, as previously stated, the most important steps are those at the beginning of the process, namely the chopping and first solvent extraction step.

In the mechanical shearing and dissolving stages, most of the gaseous and volatile fission products in the fuel will be released along with hydrogen and nitrogen oxides generated by the reaction of the metals with nitric acid. Due to the storage period prior to the initiation of the fuel reprocessing, essentially all of the short-lived radioactive gases will have decayed to stable isotopes as will the bulk of the radioactive iodine isotopes which are volatile and, biologically, highly significant. The recovery of any residual radioiodine in process vent lines is readily accomplished by reaction with silver and adsorption on charcoal. The major gas of concern is krypton-85, a long-lived noble gas (half-life of about ten years) for which decay storage is not feasible. Also some smaller amounts of shorter-lived xenon and krypton are released. In this step, tritium either may be released as a gas or may combine with the dissolver solution to form tritiated water. About 25 percent of the tritium is discharged as the gas.

The nonvolatile fission products are converted to nitrates in the dissolver along with uranium and plutonium and are passed through the solvent extraction step. Any tritium present as water also moves to the extractor. In the extractor, the uranium and plutonium nitrates are preferentially distributed into the organic extraction phase leaving the bulk of the fission products in the aqueous nitric acid solution. This acid solution is the high-level waste which is the major waste with which the reprocessing plant must cope.

LIQUID AND GASEOUS RADIOACTIVE WASTES

Radioactive liquid and gaseous wastes are accumulated in the processing steps subsequent to the first solvent extraction, however, the quantity and activity levels are substantially less than those just discussed. These subsequent waste streams are referred to as intermediate or low-level wastes and are, comparatively speaking, not troublesome. We will return to how these wastes are handled, but before we do that, let us determine how much waste might be produced.

In January, John Nassikas, Chairman of the Federal Power Commission, stated that the FPC estimate of nuclear generation capacity in the United States would reach 147,000 Mwe. or 22.1 percent by 1980 and 500,000 Mwe. or 39.4 percent by 1990. The United States Atomic Energy Commission estimates, as of 31 December 1970, that 51,535 Mwe. of nuclear generating capacity is either operable or being built. This amounts to 15 percent of the total United States capacity. In addition, an additional 35,360 Mwe. is on order. These numbers represent a projected annual increase of nuclear capacity on the order of 12–13 percent compared to an overall electrical generating capacity increase of about 7 percent annually.

In current reactor designs, average fuel life in power reactors is limited to about 25–30,000 Mwt.-days/MTU. If one assumes an average plant capacity factor of say 75 percent over the life of the plant and a thermal efficiency of 33 percent, then rough estimates of the fuel reprocessing load can be obtained. The projected annual reprocessing load for 1980 would be about 4340 MTU and for 1990 would be about 15,200 MTU.

Using these figures for the approximate annual reprocessing load, we can estimate the amount of fission product activity that must be removed from the fuel and safely disposed. Table 14.1 shows the approximate fission product activity generated in a 1000 Mw. (electrical) plant each year and the approximate amount of certain important long-lived isotopes. If we convert these numbers for what is produced in one power reactor annually to the total amount of fission products that go to the fuel reprocessing plants from all reactors, we receive the results shown in table 14.2. Figure 14.3 shows the projected cumulative total amount of fuel reprocessed from today. By 1980 approximately 29,000 metric tons will have been processed, by 1990 approximately 123,000 MTU, and by the year 2000 approximately 350,000 MTU.

At the one commercial fuel reprocessing plant in operation in the United States, the Nuclear Fuels Services plant, high-level and intermediate-level wastes are stored as liquids in underground tanks, as has been the past practice at the Atomic Energy Commission's separations facilities at Hanford, Savannah River, and Idaho Falls. The high-level wastes are those wastes which are removed as an aqueous-acid solution from the first solvent extraction step. These wastes are concentrated by evaporation to reduce the total volume of liquid waste. The concentrated material from the evaporator, which contains most of the fission products, is neutralized and stored in underground tanks. The condensed vapor from the evaporator is subjected to additional evaporation after being mixed with lower activity wastes coming from processing steps subsequent to the first extraction. Again the concentrate is stored in the underground tanks. The purified vapors are condensed and discharged through ion exchangers to a holding lagoon and ultimately discharged to surface waters. Although there may be some process differences, the Allied-Gulf Barnwell will similarly process their liquid wastes.

Significant amounts of high-level radioactive waste solutions have already been produced and stored at the AEC facilities. At the present time, over 80 million gallons of this type waste has been handled as a result of the processing operation on reactor fuels used for the production of plutonium. These wastes are being stored in special underground tanks which range in size from about 300,000 to 1.3 million gallons capacity. Although no one proposes that such storage would be a long-term solution to the problem of waste disposal, it has been demonstrated that liquid waste storage in underground tanks is satisfactory for relatively short periods (perhaps a few decades), with constant surveillance and replacement of tanks as necessary.

TABLE 14.1
1000 Megawatt Light Water Reactor Long-Lived Fission Products

Nuclide	Half Life	Curies Per Year
H-3	12.3 yr.	11,000
Kr-85	10.5 yr.	320,000
Sr-90	28 yr.	3,200,000
Cs-137	30 yr.	2,900,000

Total fission products (long-lived) $\sim$ 90,-000,000

TABLE 14.2
Total Projected Long-Lived Fission Product Activity at The Reprocessing Plant (megacuries)

Nuclide	1980	1990
H-3	1.6	5.6
Kr-85	47.5	161.9
Sr-90	475.3	1618.9
Cs-137	428.5	1459.2
Total (long-lived)	15,000	52,000

Interim or short-term storage is considered by the USAEC to be the holding of radioactive wastes safely for decades, whereas long-term storage is considered to be the containment and storage of these wastes during the hundreds or thousands of years that this material will be biologically hazardous.

At Hanford, Idaho, and Savannah River, the current practice is to convert all high-level liquid wastes into a sludge or salt cake or calcined solid, at least for temporary storage until a final decision has been made as to the method to be used for ultimate disposal. The majority of waste now stored as liquids was generated before the technology for solidification of wastes was developed.

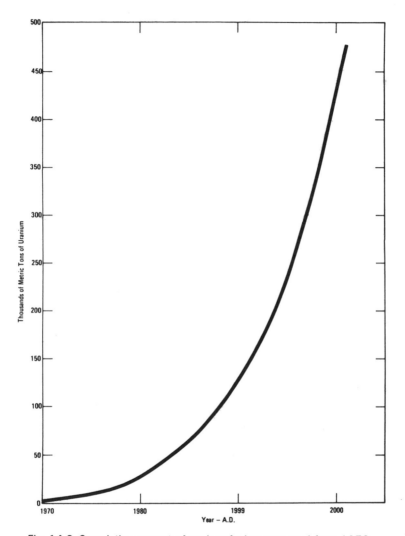

Fig. 14.3 Cumulative amount of nuclear fuel reprocessed from 1970.

Using present estimates for installed nuclear generating capacity, the volume of waste accumulated from commercial reprocessing of civilian nuclear power reactor fuels will total about 3.0 million gallons by 1980, 12 million gallons by 1990, and 45 million gallons by the year 2000, assuming all the wastes are stored as liquid at a rate of 100 gallons of waste per metric ton of uranium processed.

Research and development conducted by the Atomic Energy Commission has demonstrated the feasibility of converting these liquids to an insoluble solid form, and the resulting reduction in both volume and mobility has greatly enhanced the outlook for long-term control of these highly radioactive concentrates. Processes for the conversion of liquids to various solid forms by a variety of methods are currently under full-scale tests at AEC installations to determine the most feasible and safest technique.

The second commercial fuel reprocessing plant in the United States, the Midwest Fuel Reprocessing Plant, under construction near Morris, Illinois, will concentrate the high-level liquid waste in an evaporator and then calcine the concentrate to a solid form with an estimate of about fifteen gallons of solids remaining for each metric ton of uranium processed. On this basis, in the year 2000 only about two and one-half acres of land would be required to contain all of the solidified high-activity waste generated up to that time. However, because of the lag time between fission product generation and ultimate disposal, approximately 2000 acres of storage space will have been committed.

Considering the availability of proven techniques for solidification of high-level wastes and the notably small area required to provide permanent storage, it appears that this waste would not be a major environmental problem.

At the Midwest Fuel Reprocessing Plant, the overhead vapors from the evaporation step to concentrate liquid wastes are recycled back through the process and, consequently, are not released from the plant except perhaps in very small quantities. Low-level wastes removed in subsequent steps are either converted into fluorides, adsorbed on fused alumina and disposed of as solids, or concentrated by evaporation and the concentrate solidified by mixing with asphalt.

The Atomic Energy Commission has recently published a proposed policy on the siting and waste management facilities required for reprocessing plants. Briefly, this policy would require that high-activity liquid wastes must be converted to an approved solid form within five years and the solid shipped to an approved depository within ten years. Thus, the liquid wastes currently in temporary storage at the Nuclear Fuel Services reprocessing plant in New York will be required to be converted to solid form and shipped to federal burial sites for ultimate disposal. The primary concern of the AEC in developing this policy is the need to restrict the quantity and mobility of high-level wastes stored at reprocessing plants in the interest of the public health and safety. The Commission does not now regard storage of liquid high-level wastes in tanks as constituting an acceptable method of long-term storage.

On 17 June 1970, the Commission announced the tentative selection of a site near Lyons, Kansas, for the location of a demonstration radioactive waste repository. This repository would be constructed in a deep-bedded salt formation approximately 1000 feet underground. Such salt formations are attractive since they are found underlying an area totaling about 400,000 square miles of the United States in twenty-four states. Salt formations have good thermal and structural properties, are necessarily free of circulating ground water, and are usually completely isolated from underground aquifers by impermeable rock formations. The future of the Lyons, Kansas, repository, however, might be in doubt. The project has been attacked by at least one Kansas representative to Congress and has been questioned by other members of the Kansas congressional delegation. The repository is currently scheduled to be in operation in approximately four years.

KRYPTON-85 DISCHARGE

Assuming the availability of acceptable underground storage locations and that all high-level liquid wastes are ultimately converted to solids which are stored in these repositories in sealed containers, there is one major remaining waste management problem: What can be done to reduce the release of radioactive krypton-85, and tritium which is released either as a gas or as water vapor?

These nuclides present two somewhat different problems. The krypton-85 is by far the more important from a biological point of view. As indicated in figure 14.4, the estimated dose from uncontrolled release of krypton-85 rises rapidly as the worldwide nuclear industry develops, rapidly reaching nonhazardous, but significant, levels within the next thirty years.

As can be seen by the estimates of Coleman and Liberace (Rad Health Data and Reports vol. 7, no. 11, November 1966), the projected dose estimates assuming continued uncontrolled release of krypton-85 by about the year 2000 is 4 mrad per year. This is a dose to the whole body and is comparable to natural background whole body doses which vary from 80 mrad per year to 170 mrad per year depending upon location.

It appears that it may be technically and economically feasible to eliminate or reduce this discharge to the atmosphere. One process under development at the Oak Ridge National Laboratory utilizes the solubility of the noble gases in the halogenated refrigerants. Off-gases from the dissolver and

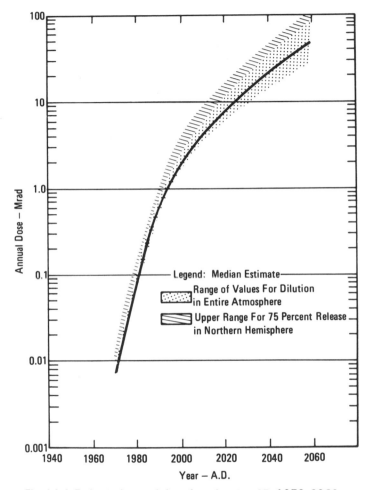

Fig. 14.4 Estimated annual dose from krypton-85, 1970–2060.

subsequent stages of reprocessing would be passed through a packed column with liquefied refrigerant flowing countercurrently as shown in figure 14.5. The off-gases, stripped of radioactive krypton and xenon, would be discharged to the atmosphere. The refrigerant leaving the bottom of the column enriched in krypton and xenon would then pass through a fractionator where any dissolved oxygen and nitrogen would be removed. The liquid would flow to a stripping column where the krypton would be removed and pumped to storage tanks for ultimate disposal. This process, still under development, promises an economical method of krypton and xenon removal.

Krypton could also be removed by a cryogenic process as shown in figure 14.6. This system basically operates to condense the krypton and xenon present in liquid nitrogen. The liquid nitrogen and condensed noble gases then flow to a distillation column where the nitrogen and undesired gases are allowed to boil off, leaving the liquefied noble gases which are then stored in pressurized tanks for disposal. This process is under development at the Idaho Chemical Processing Plant.

A third alternative, although not as attractive, is to adsorb these gases on charcoal at cryogenic temperatures, then isolate the charcoal and desorb the gases for collection and bottling for ultimate disposal.

At the present stage of development these recovery processes are expensive and since there is some lead time available, the selection of a process for incorporation into future plants and for backfitting to current plants can be deferred, pending the results of present research and development efforts underway to improve the efficiency and lower the costs of such a recovery process.

TRITIUM DISCHARGE

The other isotope, tritium, is one which presents a much greater technical problem in recovery but a much lesser biological problem. The methods currently available for the separation of tritium and normal hydrogen are extremely complex and costly and would not be feasible for use at a fuel reprocessing plant to recover tritium. However, the properties of tritium are such that the biological consequences of its release to the environment appear to be negligible. Although the basis for tritium hazard assessment has been questioned all national and international organizations concerned with radiation protection have stated that the hazard of tritium as water or vapor should be evaluated in terms of its delivered dose. As indicated in figure 14.7, the projected dose to body tissue from the release of all tritium produced from the world nuclear economy thirty years in the future is less than 0.001 mrem/year as compared to a range in natural background

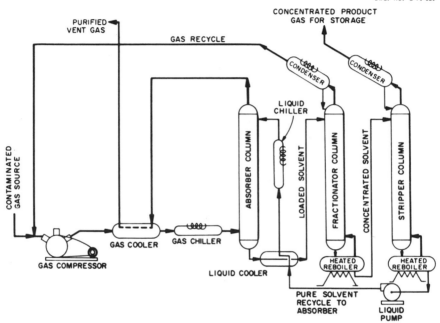

Fig. 14.5 Krypton-xenon adsorption process pilot plant schematic flow diagram.

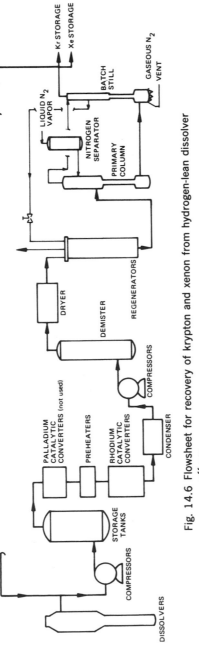

Fig. 14.6 Flowsheet for recovery of krypton and xenon from hydrogen-lean dissolver off-gas.

253

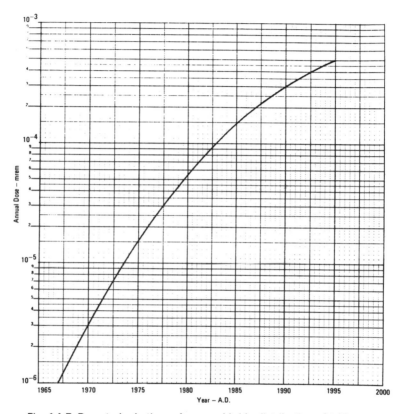

Fig. 14.7 Dose to body tissue from worldwide distribution of tritium.

radiation of between 80–170 mrem/year. This insignificant incremental dose considered on any reasonable basis should not create a biological problem of any kind.

SOLUTION TO WASTE PROBLEM

The large quantities of radioactive fission products so carefully controlled and contained in nuclear power reactors are deliberately released at the fuel reprocessing plant. The control of the so-called high-level wastes is a manageable, although expensive, process requiring essentially eternal care. The best solution at present appears to be the storage of these materials underground in geological formations which are stable and isolated from the human environment. The major environmental questions associated with reprocessing plants are those which result from the gaseous materials potentially released by such plants. Technology is currently available and is being improved for the control of krypton-85, the major potential dose contributor, and other noble gases. The projected dose increment in the future from tritium, for which no practicable recovery methods are currently feasible, does not constitute a biological problem of any magnitude. Considering all of these factors, the management of radioactive waste from fuel reprocessing does not appear to pose significant or difficult problems for solution. The technology is largely available at the present time to solve the problems that are yet in the future.

15. Present and Future Radiation Levels

John R. Totter
Director, Division of Biology and Medicine, United States Atomic
Energy Commission

The proper context for discussing environmental radiation requires a total approach involving the assessment of energy requirements, potential radiation levels, and factors both known and unknown concerning natural levels and effects of radiation. This synthesis of fact and prophecy in these areas is most relevant to the judgments regarding risk versus benefit which our society must make.

The subject of ionizing present and future radiation levels involves not only analytical and scientific measurements and calculations, but also prophecy. When prophecy is based on sound scientific and engineering principles and achievements, the element of prophecy involved with the near future takes on a semblance of probability. Society can act upon this sort of "prophecy" to guide its actions in a rational way. It can adjust its aims and choose between alternatives during the slow development of a new technology in such a way that serious risks can be avoided as new knowledge unfolds. Society and the individual can never avoid a degree of risk, however, as everything one does or fails to do involves a degree of uncertainty.

Every action which society takes or fails to take involves implicit or explicit weighing of benefits against risks. The same is true of individuals; we are accustomed to performing this kind of decision making so frequently that it is rarely a wholly conscious act.

There are prophets abroad now, a few of whom appear to be embarked on a genuine jeremiad against the development of nuclear energy. Their prophecies range from a claim that radiation exposure to the general public from nuclear power enterprises will necessarily reach maximum limits set by various official and nonofficial bodies, to claims of catastrophe from undisclosed kinds of accidents to nuclear power plants.

Let us briefly review what is known about radiation exposures from natural and artificial sources and what are estimated to be the health consequences of these exposures. Then we can attempt a "scientific" guess concerning probable exposures from current processes in development.

NATURAL SOURCES OF RADIATION

Major natural sources of ionizing radiation which reach us are listed in table 15.1. Radiation from these sources is particulate: electrons, neutrons, mesons, helium nuclei, etc.; and electromagnetic: x-rays and gamma rays.

TABLE 15.1
Natural Sources of Radiation

1. Cosmic rays	
2. Radioisotopes produced by cosmic rays	(H^3, C^{14}, Be^7, etc.)
3. Other radioisotopes	(U^{238}, Th^{232}, K^{40}, Ra^{226}, etc.)

Cosmic rays are atomic nuclei with energies ranging up to at least 10^{20} electron volts, which arrive at the top of the earth's atmosphere at the rate of a few particles per square centimeter per second. Degradation of their energy upon interaction with the atmosphere produces a flux of γ-rays, charged particles and neutrons. At sea level the charged particles and γ-rays deposit approximately 30 millirads/yr. and the neutrons about 1 millirad/yr. at our latitudes. Fluxes and doses increase about twofold for every two kilometers of altitude near sea level. There is a latitude dependence because the earth's magnetic field selectively protects the equatorial region where doses are some 10 percent lower than at mid-latitudes. Discounting minor short-term variations, the flux of incident cosmic rays has apparently been constant for many thousands of years.

A number of relatively short-lived isotopes are produced and maintained at steady state amounts by cosmic ray interactions with the atmosphere.

Tritium is produced at a rate of about 5 megacuries per year, which maintains a natural inventory in excess of 70 megacuries. Highest concentrations of this tritium are found in atmospheric hydrogen and water vapor. Water in the deep ocean and in deep wells contains very low concentrations of tritium.

Carbon-14 is produced at the rate of about 25,000 curies per year, maintaining a world-wide inventory of about 250 megacuries against decay with a half-life of nearly 6,000 years. Only about three percent of this soft beta emitting isotope is found in biological materials. Nearly 95 percent is in the oceans with nearly all of this found below the thermocline.

These and other isotopes produced as a result of cosmic ray interactions are listed in table 15.2.

TABLE 15.2
Some Radioisotopes Produced by Cosmic Rays

Isotope	Half-life	Concentration (dis./min./cu.m.) *
Tritium-3	12.3 yrs.	10^1
Beryllium-7	53 days	1
Beryllium-10	2.7×10^6 yrs.	10^{-7}
Carbon-14	5760 yrs.	4
Sodium-22	2.6 yrs.	10^{-4}
Silicon-32	700 yrs.	2×10^{-6}
Phosphorus-32	14.3 days	2×10^{-2}
Phosphorus-33	25 days	1.5×10^{-2}
Sulfur-35	87 days	1.5×10^{-2}
Chlorine-36	3×10^5 yrs.	3×10^{-8}

SOURCE: "The Natural Radiation Environment," *Understanding the Atom*, USAEC Division of Technical Information (1968).
* Disintegrations per minute per cubic meter of air in the lower troposphere.

Very long-lived radioisotopes, which have been present since the formation of the earth, and their radioactive decay products are important sources of radiation. Some of these radioisotopes are listed in table 15.3.

TABLE 15.3
Radioactive Elements in the Lithosphere

Isotope	Abundance	Half-life (years)	Radiation
Radium-226	2×10^{-12} g/g *	1622	alpha, gamma
Uranium-238	4×10^{-6} g/g	4.5×10^9	alpha
Thorium-232	12×10^{-6} g/g	1.4×10^{10}	alpha, gamma
Potassium-40	3×10^{-4} g/g	1.3×10^9	beta, gamma
Vanadium-50	0.2 ppm. †	5×10^{14}	gamma
Rubidium-87	75 ppm.	4.7×10^{10}	beta
Indium-115	0.1 ppm.	6×10^{14}	beta
Lanthanum-138	0.01 ppm.	1.1×10^{11}	beta, gamma
Samarium-147	1 ppm.	1.2×10^{11}	alpha
Lutetium-176	0.01 ppm.	2.1×10^{10}	beta, gamma

SOURCE: "The Natural Radiation Environment," *Understanding the Atom*, USAEC Division of Technical Information (1968).
* Gram per gram of soil.
† Parts per million.

Several members of the uranium series (uranium-238 and its decay products) are important natural sources of environmental radiation. This series has a reasonably long-lived gaseous member, radon-222, some of which escapes from the soil at an annual rate of about 15,000 megacuries per year and decays to solid descendants. The natural inventory of radon-222 is about 25 megacuries. Airborne members of this series contribute to lung dose and to external dose. Radon-222 and its daughters are of greater significance indoors (or in mines) than outside.

The thorium series is generally similar to the uranium series, but the short half-life of radon-224 limits airborne daughters, and the major environmental dose from this series is due to γ-rays from members of the series contained in soil and rocks.

These natural sources produce a background dose rate which varies from place to place, but varies relatively little with time (see table 15.4).

TABLE 15.4
The Natural Background

Type	Source	Dose rate in millirads/yr.	Varies with
External	Cosmic rays	30–60	Latitude and altitude
	Soil potassium-40, thorium, uranium	30–100	Location (mineral deposits) and dwelling (least in tents, greatest in stone buildings)
Internal	Thorium, uranium and daughters	40–400	Location and water supply
	Potassium-40	20	Not very variable
	Carbon-14	2	Not very variable
	Tritium	0.04	Not very variable
Total		100–600	

SOURCE: "Your Body and Radiation," *Understanding the Atom*, USAEC Division of Technical Information (1967).

The geographical variation is not very large. It is unusual to find levels less than 25 or greater than 100 millirads/yr. for terrestrial γ-rays in populated areas of the United States. Higher levels occur at rock outcrops, but these occur only rarely in populated areas. This and a number of other factors tend to homogenize the distribution of radiation sources which influence population exposure.

ARTIFICIAL SOURCES OF RADIATION

In addition to natural sources, a number of radiations come from an increasing variety of man-made sources. Some of these are listed in table

15.5. The doses and dose rates produced by some of these sources are listed in table 15.6.

TABLE 15.5
Radiations from Artificial Sources

Radiation	Source
Gamma rays	Radioisotopes, reactors, accelerators, etc.
x-rays	x-ray machines used for medical and industrial purposes
Beta rays	Radioisotopes (fallout and the nuclear industry)
Neutrons and other particles	Reactors, accelerators, spontaneous fission of some heavy radioisotopes

TABLE 15.6
Nonoccupational Artificial Exposures

Source	Dose or Rate
Wrist watch dial, approximately 1 microgram of radium, gamma rays	1 mr./hr.
Airplane instruments—pilot position	1 mr./hr.
Shoe fitting (20 sec.)	10 r.
Diagnostic x-ray	
14 × 17 chest plate	0.1 r.
Photofluorographic chest	1 r.
Extremities	0.5 r.
GI series (per plate)	1 r.
Pregnancy	9 r.
Fluoroscopy	15 r./min.
Dental (per film)	0.5 r.

SOURCE: "The Natural Radiation Environment," *Understanding the Atom*, USAEC Division of Technical Information (1968).

There are variations in natural background, but they are not large enough to make it possible to detect differences in human populations. All existing human data result from relatively small populations which have received high levels of exposure from artificial sources (atom bombs, x-ray machines) or were deliberately exposed to high levels of natural sources (radium dial painters, thorotrast patients). Data acquired on these latter groups make it very clear that it will be difficult if not impossible to observe differences between population groups exposed at even the extreme ranges of natural background.

One of the highest background areas is in the Indian state of Kerala. Perhaps 100,000 people in this area have been exposed to dose rates which range from two to ten times the average background rate. However, epidemiological information on the population is meagre and a control population for comparison is difficult to delineate.

The 50,000 or so people in the Guarapari area of Brazil may be too small a population to provide suitable statistics.

The Massif Central area of France has a population of about 7,000,000. It is said that background exposure of this group is about double the exposure in the remainder of France. The four departments having the highest background exposures, however, have relatively few people.

It has been estimated that a population of 6,000,000 would have to be observed for a year in order to demonstrate the effect of 5 rads on the incidence of leukemia.[1] As a result there is no direct way in which to determine accurately the consequences to humans of irradiation at the low chronic levels to which we are constantly exposed. Estimates of effects at these or lower levels must be made by observing the effects produced by high doses given at high dose rates in animal or human populations, and then extrapolating to low doses and rates according to some assumed mathematical relationship.

At high doses there are two types of effects which are differentiated. These are genetic effects which appear only in the descendants of the irradiated person and somatic effects which occur in the irradiated individual but are not transmitted to his descendants. Quantitative information on genetic effects of irradiation in mammals is known only from animal experiments. Quantitative data on somatic effects are available from both human exposures and animal experiments.

Mouse genetic experiments conducted chiefly by Russell and co-workers on human incidence of genetically related diseases has led to an estimate of a doubling dose of about 50 rem in humans.[2] Thus, an acute whole body dose of 50 rem of ionizing radiation to a large population of humans might lead to about twice as many mutations in the germ cells of this population as would have occurred naturally. The average generation time is taken as thirty years; a person exposed continuously at the maximum exposure limit would accumulate 5 rem in thirty years. Assuming that the mutation rate increases linearly with dose, this would produce an increase of ten percent in the naturally occurring mutation rate. In this estimate, no allowance has been made for a possible sparing effect of the very low dose rate at which background radiation is delivered. In experiments with mice at low dose rates (still very much higher than background rates) estimates of the mutation rate were about sixfold lower than at high dose rates. The doubling dose for mutations, therefore, with protracted or low rates of exposure may be as high as 300 rem.

It is believed that about 2–4 percent of newborn children have genetically related defects. Diseases which are genetically related through the operation of a large number of genes may include a significant proportion of noninfectious maladies, such as diabetes, hypertension, and others from which humans suffer. There is little or no conclusive information as to how the manifestations of these diseases are controlled genetically (i.e., single or multiple gene control) or to what degree the final expression of disease may be affected by environmental factors.

Somatic effects of radiation have been extensively studied in humans. The chief groups involved have been persons exposed to radium from dial painting or from medical uses; survivors of the Hiroshima and Nagasaki bombings; radiologists; and patients who have been treated therapeutically by x-rays.

The radium dial painters and related groups have provided information on the induction of osteogenic sarcomas from bone-seeking radioisotopes. Long-term comparative experiments with animals are making it possible to relate this quantitative information to the effects of man-made bone-seeking alpha and beta emitters in man.

The atomic bomb survivors appear to afford the best information for the effects of external radiation. The most definite quantitative data appear to be those on leukemia induction. The relationship is often stated as 1–2 extra leukemias per year per million man rems (man rems = number of persons exposed times dose in rems). After twenty-five years, the actual count is approximately thirty excess leukemia cases per million person rems and the early high rate of occurrence seems to have tapered off. Other human cancers which appear to be induced by whole body radiation are thyroid adenomas, and possibly cancer of the lung and tumors of the breast. Many types of cancer have not yet shown an increase in incidence in the exposed Japanese population. Jablon and Kato report little or no increase in the childhood leukemias or cancer resulting from *in utero* exposure.[3]

At the present time accurate statistics on the differences in total cancers in the exposed and nonexposed groups are not yet available. Information published by the scientists of the Atomic Bomb Casualty Commission, however, makes it possible to estimate that the ratio of total excess cancers to excess leukemia in the exposed group is not greater than approximately four, and it may be much less.

Estimates of this ratio by Gofman and Tamplin vary from approximately two to thirty-one.[4] The high estimates appear to be partly based on projections of incidence rates into the future for the Japanese. The majority of the populations at risk used in these projections are still surviving; and it will be several years before final conclusions can be drawn.

One study is complete, however, and cannot change since all members of the study group are now dead. This is the group of radiologists studied by Warren,[5] and by Seltser and Sartwell.[6] Unfortunately, the doses received by individuals in this group are not known and, as a result, have been variously estimated as being in the range of several thousand rads.

The number of "extra" cancer cases was not quite equal to the extra leukemias. However, in part, because the average life span of the radiologists was some five years shorter, the age specific incidence of cancer was appreciably higher than in the controls. It is instructive to examine this more closely.

In a numerically static, closed population the birth rate and death rate are equal. If this equilibrium is disturbed by introducing an agent which increases the death rate, the population will tend toward a new equilibrium at which the former death rate will be restored. The population will then differ by being somewhat smaller in size, having a shorter average life span, and the age specific death rate will be somewhat elevated. There will be "extra deaths" only during the transient period before reaching the new equilibrium.

In the radiologist group all major causes of death had a higher age specific incidence. Therefore, most deaths contributed to the shorter average life span. Perhaps only the leukemias contributed much more than their usual share of influence on the average age at death. If this is true, then one can say with some confidence that the most proper measure of the somatic effects of small doses of ionizing radiation on a population is its effect on the life span. Making use of the estimate given above for the radiologists, 5 rads in thirty years would tend to reduce the average life span by $5/2000 \times 5 \times 365 = 5$ days.

The radiologist studies are important in pointing out to us that the major cancerogenic effect of small but significant exposures of a population to ionizing radiation is likely to be not "extra" cancers but approximately the same numbers of cancers occurring a few days or weeks earlier than they would otherwise have occurred anyway. If one takes into account the probable reduction in effect owing to slow delivery rate, the effect indeed may be very very small.

After this short and necessarily sketchy review of radiation effects, let us consider projections for population exposures in the future.

FUTURE PROJECTIONS FOR RADIATION EXPOSURES

Approximately 90 percent of dose to man (man-rads) is now being delivered by natural sources of radiation. My most confident prediction regarding the future is that this percentage will change very little over the next three decades. My estimates of dose to man are shown in table 15.7; the current man-rad exposure rate from natural sources is projected to double over the next thirty years, on the basis that dose rates from natural sources remain constant while the world's population increases twofold.

TABLE 15.7
Estimates of Dose To Man

	Man Rads Per Year	
	1970	2000
1) Natural sources	$4 \cdot 10^8$	$8 \cdot 10^8$
2) Medical exposures	$2 \cdot 10^7$	$5 \cdot 10^7$
3) Fallout	$3 \cdot 10^6$	$6 \cdot 10^6$
4) Occupational exposure	$2 \cdot 10^5$	$5 \cdot 10^5$
5) Tritium	$2 \cdot 10^5$	$4 \cdot 10^5$
6) Krypton-85	$3 \cdot 10^3$	$2 \cdot 10^5$
7) Airline travel	$2 \cdot 10^3$	$1 \cdot 10^5$

Any change that does occur is likely to be associated with medical uses of radiations and radioactivity, which now account for most of the remaining dose to man. All evidence seems to indicate that the rapidly increasing per capita use of medical procedures using radiation and radioisotopes will continue for a number of years. It also appears probable that these procedures will become available to an ever-increasing fraction of the world's population. This could lead to an increase in the man-rad exposure rate many times larger than the entire increase likely to arise from all other man-made sources.

It is very likely, however, that there will continue to be improvement in techniques and equipment for reducing dose in these procedures, and that this will permit the benefits of greater use with less than proportionate increase in dose. Thus, in table 15.7, the contribution from medical sources in the year 2000 has been projected by anticipating a factor of two reduction in dose per procedure. This anticipated improvement reduces the projected rate by 5×10^7 man-rads per year, a reduction in itself substantially larger than all other sources of man-made radiation. There is, therefore, reason to expect large returns in dose reduction per dollar expended in this area.

Currently, the majority of the dose to man from artificial radioactivity comes from fallout radionuclides, and the largest contributor to whole body exposure is cesium-137. This is included in table 15.7. A significantly larger dose, which is deposited chiefly in bone, comes from strontium-90. Since it chiefly irradiates the skeleton, it has not been included in the comparison. Projecting these contributions into the future requires prophecy. If there is no further weapons testing in the atmosphere, the dose rate will fall rapidly and its contribution to the man-rad dose rate will become very small by the year 2000. I hope this is indeed the case; however a constant dose rate from this source that may be maintained by a moderate rate of weapons testing in the atmosphere by foreign nations has been projected.

Whereas, in the case of the fallout contribution, the current estimates available cannot be projected reliably; in the case of occupational exposure, projection is reliable but little knowledge of the current value is known. It is unlikely that the value included in table 15.7 is too low; it may, however, be high by more than a factor of ten. Much uncertainty concerning this number exists because most of the dose received by individual workers is unmeasurably small. Whatever the current occupational exposure rate, it is likely to increase as an increasing fraction of the population is occupied in jobs related to use of radiation and radioisotopes.

The contribution listed for tritium is that associated with the present level of tritium in the biosphere. Nuclear operations, mostly reactor power stations and fuel reprocessing plants, will release large amounts of tritium, but at a rate which is unlikely to exceed the rate at which tritium is now being lost from the biosphere. Levels in the biosphere are now higher than natural levels as a result of weapons testing. Except for specific situations where local concentrations might become significant, it would appear that one can expect relatively small cost benefit from control of these tritium releases.

The other radioisotope produced in nuclear operations and released to the environment in substantial quantities is krypton-85. Very nearly all of this isotope that has been produced has been released to the environment. I have included the whole body dose it produces in table 15.7. Since this isotope is predominantly a beta emitter, it produces a skin dose which is some fifty times larger; this skin dose does not appear in the comparison.

In contrast to tritium, if all krypton-85 produced in nuclear operations continues to be released to the environment, we will reach concentrations in the atmosphere which are very substantially above current levels by the year 2000. This would produce a dose contribution from krypton exceeding that projected for tritium.

Procedures for removing and containing krypton and other noble gases have been developed and are working well. The figure projected for krypton anticipates some use of these removal systems. How extensively these are employed and how much krypton will be removed will be influenced by international agreements and arrangements as well as by economics and local considerations. The figure chosen is largely arbitrary but is intended to indicate that a removal capability exists which is expected to be used to some extent.

Finally, figures for dose rates associated with airline travel are given. Since aircraft which fly faster, fly higher, the dose per passenger mile has not changed very much in passing from props to jets and will not change much for the SST. Therefore, the year 2000 figure was computed by projecting a 15 percent annual increase in passenger miles over thirty years to arrive at a man-rad contribution on the same order as that for tritium and krypton.

I believe these projections and our current knowledge of radiation effects clearly indicate that we will be able to derive substantial benefit from radiation and radioisotopes without unacceptable hazard, if we are not stampeded by alarmists into irrational and irresponsible actions.

FUTURE RADIATION LEVELS

In reference to present and future radiation levels, it is clear that the best decisions, directions, and rates of progress of our use of nuclear fuels depend on actions involving the relationships between energy and the environment, not the least of which is the educational task of communicating to all citizens the benefits to be derived and the risks to be taken. We will never convince those few who will not listen, and we will never be able to promise zero risk. We can, however, provide energy, protect man, improve the environment, communicate the costs and alternatives, and move forward to a fuller life for all men, present and future.

Reference notes

1. C. Buck, "Population Size Required for Investigating Threshold Dose in Radiation-induced Leukemia," *Science* 129 (1959): 1357.
2. W. L. Russell, "Mutagenesis in the Mouse and Its Application to the Estimation of the Genetic Hazards of Radiation" (Proceedings of the 4th International Congress of Radiation Research, Evian, France, 29 June–4 July 1970).
3. S. Jablon and H. Kato, "Childhood Cancer in Relation to Prenatal Exposure to Atomic Bomb Radiation," *Lancet,* ii (1970): 1000.
4. J. Gofman and A. Tamplin, "The Cancer-Leukemia Risk from FRC Guideline Radiation Based upon ICRP Publications" (GT-117-70) in *Environmental Effects of Producing Electric Power,* pt. 2, vol. 2 (Hearings before the Joint Committee on Atomic Energy, January–February 1970), pp. 2169–79.
5. S. Warren, "Longevity and Causes of Death from Irradiation in Physicians," *Journal American Medical Association* 162 (1956): 464.
6. R. Seltser and P. Sartwell, "Ionizing Radiation and Longevity of Physicians," *Journal American Medical Association* 166 (1958): 585.

16. Development of Federal and State Air Quality Control Programs

J. E. Norco
Argonne National Laboratory

Prior to federal air quality control legislation, air pollution programs were generated on a random basis almost always at the local level in response to particular problems, such as vehicular emissions in the Los Angeles area. Comprehensive statewide programs were scarce and were, in many cases, in conflict with local agencies. Enactment of the Clean Air Act indicated the need for comprehensive control programs for air pollution. The federal government has become the driving force behind the development of state programs, and states that would not ordinarily have developed these programs are now required to fulfill certain requirements.

This chapter will deal with two major topics. The first will involve a brief delineation of the implications of the Clean Air Act of 1967 and its 1970 Amendments, with emphasis on the required responses at the state level through the implementation planning procedure. The second topic will deal with the fundamental components of an effective air quality control program.

IMPLICATIONS OF THE FEDERAL LEGISLATION

Under the Clean Air Act of 1967, the concept of the implementation planning procedure was established. It required that the states submit comprehensive plans for programs to control air pollution within air quality control regions or air sheds which were designated by the Department of Health, Education, and Welfare. Figure 16.1 shows the flow diagram for

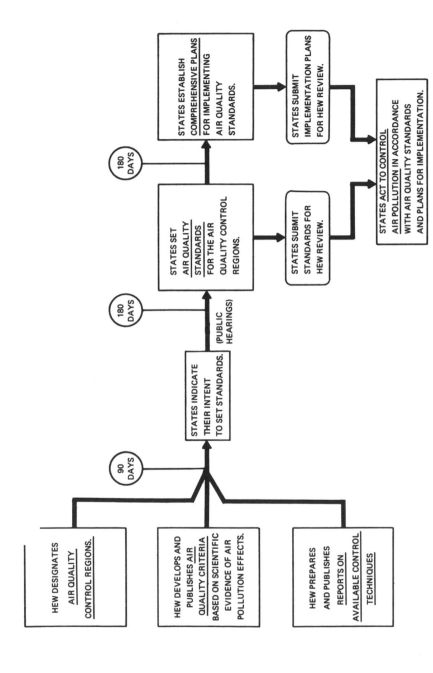

Fig. 16.1 Flow diagram for action to control air pollution on a regional basis under the Air Quality Act.

action to control air pollution on a regional basis under the 1967 Act. Under the Act, the states were required to submit separate plans for the control of sulfur dioxide and particulates for each region. The vehicular pollutants were not alluded to in the 1967 Clean Air Act. Air quality standards were set up by the states subject to federal approval, based on published air quality criteria documents which provide scientific evidence of air pollution effects on health, materials, and plants. HEW also published reports on available control techniques which were used for the preparation of emission control regulations. As shown in figure 16.1, a specific timetable of events was required for each stage of the implementation planning procedure. This meant that implementation plans were received by the federal government at a somewhat intermittent rate, depending on how the air quality regions were designated and how rapidly the states were able to respond to the requirements.

In December of 1970, the Clean Air Act Amendments were passed. The Amendments rectified many problems which were brought out in the three years after the Clean Air Act of 1967 was enacted. The 1970 Amendments have provided for a more practical and effective approach with a decreased emphasis on such activities as regional designations, extensive publication of criteria, detailed studies of the relationship of criteria to ambient standards, and the development of sophisticated diffusion models to relate ambient standards and emission standards. From the states' point of view, the 1970 Amendments represent a significant change in the game plan for the implementation planning procedure. Figure 16.2 schematically represents the revised schedule of the sequence of events for submission of implementation plans under the amended Act. Among the more important differences between the 1970 Amendments and the 1967 Act are:

1. The implementation plan for each state must include all regions or portions thereof within its boundaries.
2. The plans must provide for the achievement of the federal air quality standards (fig. 16.3) or state standards, which are at least as stringent as the federal standards.
3. The plans must provide for the abatement procedures for the first pollutant set (sulfur dioxide and particulates) as well as the second pollutant set (carbon monoxide, hydrocarbons, nitrogen oxides, and oxidants).
4. All the plans are due at the same time, 1 February 1972.
5. The air quality regions are classified according to the severity of the pollution for each pollutant set on three levels; priorities 1, 2, and 3 and the abatement procedures are to be developed accordingly.
6. The federal authority is more clearly defined. The federal government

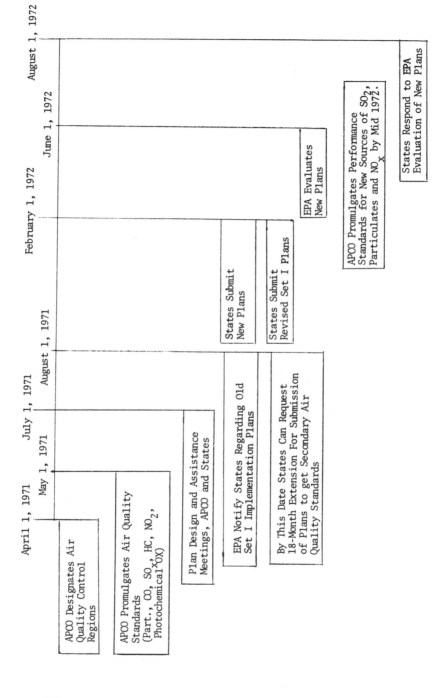

Fig. 16.2 Schedule of federal and state activities according to Clean Air Act as Amended December 1970.

TABLE 16.1—National Ambient Air Quality Standards

1. Sulfur oxides as sulfur dioxide

   Primary standards
   - a) 80 $\mu g/m^3$ annual arithmetic mean
   - b) 365 $\mu g/m^3$ maximum 24-hour concentration not to be exceeded more than once per year

   Secondary standards
   - a) 60 $\mu g/m^3$ annual arithmetic mean
   - b) 260 $\mu g/m^3$ maximum 24-hour concentration not to be exceeded more than once per year
   - c) 1300 $\mu g/m^3$ maximum 3-hour concentration not to be exceeded more than once per year

2. Particulate matter

   Primary standards
   - a) 75 $\mu g/m^3$ annual geometric mean
   - b) 260 $\mu g/m^3$ maximum 24-hour concentration not to be exceeded more than once per year

   Secondary standards
   - a) 60 $\mu g/m^3$ annual geometric mean
   - b) 150 $\mu g/m^3$ maximum 24-hour concentration not to be exceeded more than once per year

3. Carbon monoxide

   Primary and secondary standards
   - a) 10 mg/m^3 maximum 8-hour concentration not to be exceeded more than once per year
   - b) 15 mg/m^3 maximum 1-hour concentration not to be exceeded more than once per year

4. Photochemical oxidants

   Primary and secondary standards

   125 $\mu g/m^3$ maximum 1-hour concentration not to be exceeded more than once per year

5. Hydrocarbons †

   Primary and secondary standards

   125 $\mu g/m^3$ maximum 3-hour concentration (6 to 9 A.M.) not to be exceeded more than once per year

6. Nitrogen dioxide

   Primary and secondary standards
   - a) 100 $\mu g/m^3$ annual arithmetic mean
   - b) 250 $\mu g/m^3$ 24-hour concentration not to be exceeded more than once per year

SOURCE: "National Primary and Secondary Ambient Air Quality Standards," *Federal Register* 36, no. 84 (30 April 1971).

can prosecute individual polluters or take over the implementation plan after thirty days if the states fail to do so.

The Air Pollution Control Office of the Federal EPA has published a requirements document which outlines the steps and lays out the required components for the implementation plans.[1] A description of these components will follow.

COMPONENTS OF A COMPREHENSIVE STATEWIDE AIR QUALITY CONTROL PROGRAM

Legal Authority

The following seven points of legal authority are deemed necessary for an effective control program:

1. The authority to adopt emission standards and limitations, such as sulfur content laws and particulate emission regulations.
2. The authority to enforce applicable laws including the authority to seek injunctive relief.
3. The authority to abate pollutant emissions on an emergency basis, i.e., episode control regulations.
4. The authority to establish and operate a statewide system under which permits would be required for the construction and operation of pollution sources.
5. The authority to obtain information necessary to determine whether sources are in violation, such as emission inventory data.
6. The authority to require owners and operators of stationary sources to install stack testing or monitoring devices on which the data must be made available to the control agency.
7. The authority to carry on a program of inspection and testing of motor vehicles.

Source Surveillance and Enforcement

Source surveillance is generally considered to consist of two major areas, air quality monitoring and emission monitoring.

Air quality surveillance within a region must provide information to be used as a basis for the following actions:

1. To judge compliance with and/or progress made toward meeting ambient air quality standards.
2. To activate emergency control procedures to prevent air pollution episodes.

3. To observe pollution trends throughout the region including the nonurban areas. (Information on the nonurban areas is needed to evaluate whether air quality in the cleaner portions of the region is significantly deteriorating.)
4. To provide a data base for application and evaluation of effects, urban land use and transportation planning, development of abatement strategies, and development and validation of diffusion models.

An air quality surveillance program is composed of three distinct but interrelated elements: networks, laboratory support, and data acquisition and analysis. Knowledge of the existing pollution levels and patterns within the region is essential in network design. The areas of highest pollution levels must be defined together with geographical and temporal variations in the ambient levels. Isopleth maps of the ambient concentrations derived from past sampling efforts and/or diffusion modeling are the best tools for determining the number of stations needed and for suggesting the station locations. Figure 16.3 provides a first approximation of the number of stations required in a region; the number of stations is shown as a function of total population. Although the curves provide good estimates for application to population related pollutants usually from motor vehicles, such as CO, hydrocarbons, nitrogen oxides, and oxidants, they do not necessarily apply as well to SO_2 and particulate matter. For the latter pollutants, industrial complexities and fuel use patterns in the region strongly influence the pollution levels, and, thus, affect network size regardless of population. Figure 16.3 also indicates the mix of mechanical samplers for longer-term averages and automatic or continuous samplers for short-term averaging and episode monitoring.

The effectiveness of air pollution control regulations and their ultimate impact upon the air quality depends critically upon enforcement. As with the design of any enforcement system, the air pollution control enforcement must provide:

1. A clear understanding by the pollution source operators of the interpretation and implication of the regulations
2. A technically oriented set of criteria by which an operator of a pollution source knows that he is in compliance with the law
3. A set of procedures by which the air pollution agency and the pollution source operator can communicate
4. A clear statement of the information necessary from the source operator in order to determine compliance with the regulations
5. The necessary type and level of manpower to ensure that an air pollution control agency can properly carry out its enforcement functions

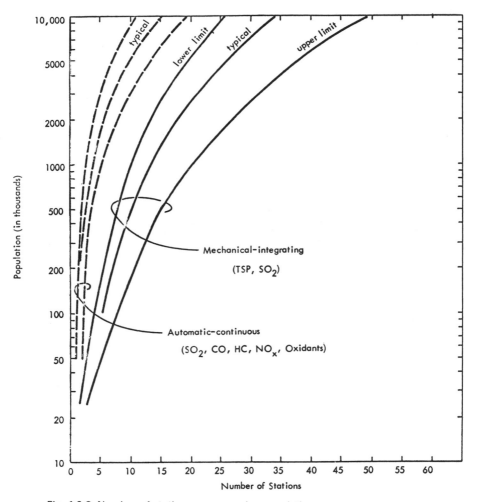

Fig. 16.3 Number of stations versus region population.
SOURCE: *Guidelines: Air Quality Surveillance Networks,* Division of Air Quality
and Emission Data, Air Pollution Control Office, Environmental Protection
Agency (17 December 1970).

Regardless of the type of regulation, there is a small number of surveillance techniques which can be used individually or in combination in the enforcement system. The first of these is the *emission inventory*. The emission inventory is utilized to keep track of all pollution sources and control devices usually by means of a mail-out questionnaire. This information (together with standard emission factors) is used to estimate the amounts of pollution being generated and its spatial distribution. Although the emission inventory is an excellent technique for keeping general surveillance of the emission patterns, it is generally conceded that it is very difficult to keep an emission inventory updated and that regulations regarding the reporting of pollution sources and control devices have not been stringent enough to ensure data that is detailed enough for enforcement purposes. Emission inventories necessitate the use of efficient data processing techniques for updating and report generation.

A second method of source surveillance is the *engineering evaluation* of plans to install or modify pollution sources and control devices (see fig. 16.4). This usually entails a manufacturer sending the plans for new pollution sources, with their plans for pollution abatement, to the air pollution control agency prior to their construction. A pollution control engineer then reviews these plans. A qualified engineer is necessary for this evaluation since it requires the knowledge of design specifications and engineering criteria of pollution control devices.

A third type of surveillance technique is the *engineering inspection* (see fig. 16.5). This is generally an on-site inspection made by a technician-class employee. The purpose of this inspection is to ensure that the information on an emission inventory form is correct; that is, that all pollution sources have indeed been registered and that the control devices that are specified on the emission inventory form are actually in operation. The inspector also checks to see that these control devices and pollution sources are being maintained properly so that the stated efficiency of the control device is accurate. Engineering inspections usually do not generate any more information than is found on an emission inventory form, but are used primarily as a method of insuring that the emission inventory form has been filled out correctly and completely. It is the information on the emission inventory form which must indicate whether or not the source is in compliance with the regulation.

The fourth type of surveillance involves *direct testing* of the concentrations of pollutant by stack testing and laboratory analysis of fuels. This testing requires that a team of two or three inspectors take air samples from the pollution source. Stack tests and laboratory analyses are probably necessary to produce sufficient evidence in any major type of violations, hearings, or major appeals.

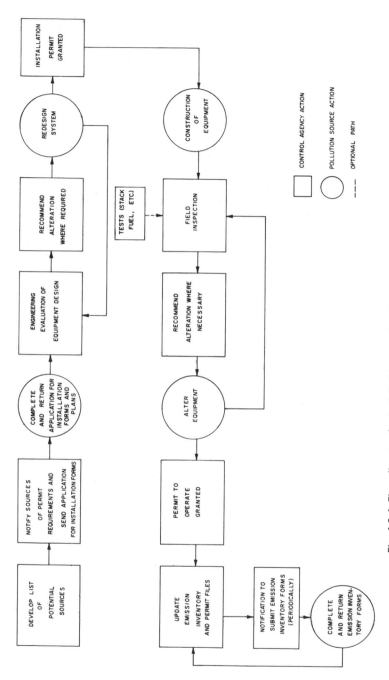

Fig. 16.4 Flow diagram for installation permit system.
Source: K. G. Croke and J. E. Norco, "An Evaluation of Alternative Source Surveillance Systems," Air Pollution Control Administration Paper no. 71-8, Air Pollution Control Administration Conference (Atlantic City, N.J., June 1971).

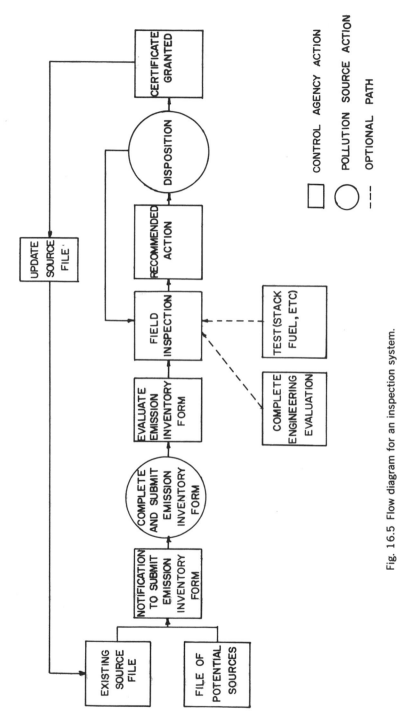

Fig. 16.5 Flow diagram for an inspection system.
Source: K. G. Croke and J. E. Norco, "An Evaluation of Alternative Source Surveillance Systems," Air Pollution Control Administration Paper no. 71-8, Air Pollution Control Administration Conference (Atlantic City, N.J., June 1971).

279

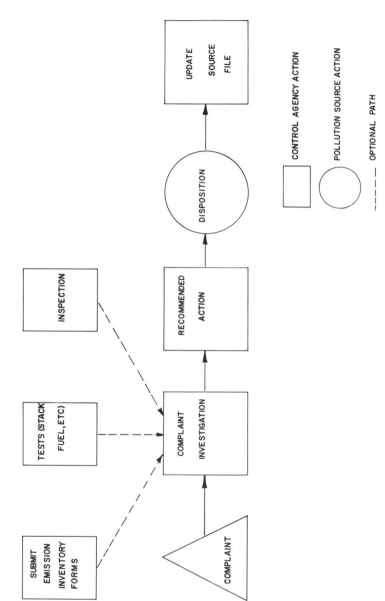

Fig. 16.6 Flow diagram for complaint system.
Source: K. G. Croke and J. E. Norco, "An Evaluation of Alternative Source Surveillance Systems," Air Pollution Control Administration Paper no. 71-8, Air Pollution Control Administration Conference (Atlantic City, N.J., June 1971).

A fifth method of surveillance is the *complaint inspection* (see fig. 16.6). In this method, only when a complaint or series of complaints is sent to the air pollution agency is an inspector sent to inspect a source of pollution. This must usually be done on a real-time basis so that the complaint inspector can survey the pollution source as quickly as possible after the complaints have been made. This is necessary both in order to have an eye-witness of the event and to indicate to the people that the agency is enforcing the regulations promptly and thoroughly. The difficulty with this type of enforcement is that it is frequently difficult to relate air contamination with a particular source, especially in the case of nonvisible emissions. Thus, the complaint form and complaint inspection can be effective with respect to the more visible emissions, but cannot be relied upon completely to enforce the full intent of the regulations.

Episode Control System

Air pollution planning is generally applied on two time scales: the long-term (annual) averages, which are attained through emission control regulations such as sulfur content and particulate limitations, and the short-term episode situations, which are usually associated with stagnating meteorological conditions causing short-term peaks on the order of a few hours to several days. Severe adverse health effects have been observed during these periods of high pollutant buildup; an efficient system for coping with the episode situation is extremely important. The essential elements of an effective episode system are given below:

1. A set of episode alert criteria based on pollutant concentrations for the initiation and transition through a multi-stage episode control plan such as that shown in figure 16.7. These are written so that more severe pollutant concentrations will trigger progressively more stringent control actions.
2. A set of preplanned episode control strategies must be established. For example, diffusion model results such as isopleth maps can be precalculated for various expected episode meteorological situations. Based on these results and logbooks of dual fuel sources, preplanned strategies can be developed for achieving immediate reduction of air contaminant emissions which are specific for each alert level and cover the entire area of emissions including fuel combustion, industrial processes, motor vehicle emissions, and solid waste disposal. With this information upwind sources can be controlled for a certain meteorological set of conditions so as to get the most benefit with a given and usually limited allocation of clean fuels.
3. An emergency operations control center must be established. This

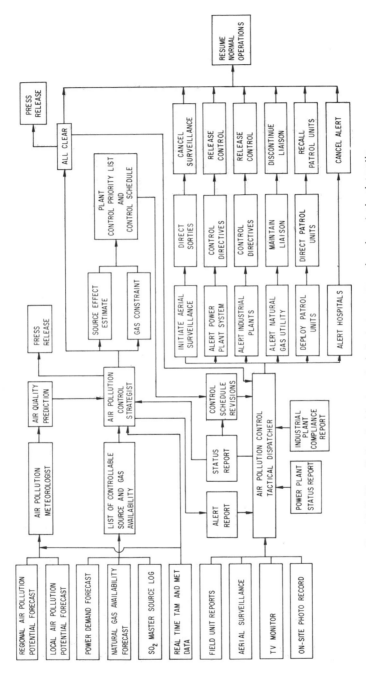

Fig. 16.7 Air pollution incident control strategy command and control schematic.
Source: E. J. Croke et al., *Air Pollution Incident Control Manual*, Argonne National Laboratory (to be published).

control center is the hub of communications for the coordination of the control activities between all municipal, county, and state control agencies.

4. A well planned information-dissemination program serviced by an efficient communications system is essential to the smooth operation of an episode control plan. The effectiveness of the system is highly dependent upon rapid and accurate transmission of (1) input data from surveillance equipment to the control center and (2) abatement instructions to the various sources. Communications and procedures manuals are required to clearly delineate the activities of the personnel through the various stages of the alert plan.

Reference notes

1. "Requirements for Preparation, Adoption, and Submission of Implementation Plans," *Federal Register* 36, no. 67 (7 April 1971).

17. Environmental Engineering for the Navajo Project

T. M. Morong
Salt River Project
J. L. Shapiro
Bechtel Corporation

The Navajo Project is an outgrowth of meetings initiated in 1968 between a group of organizations interested in additional electric generating capacity in the Southwest. As a result, a joint study committee undertook an intensive investigation of alternatives involved in meeting predicted electrical demands. Its conclusions formed the basis for the decision of the Navajo Project participants to proceed with construction of a large coal-fired generating station near Page, Arizona. A list of the participants, with their percentage of participation and utilization of power, is shown in table 17.1.

As can be seen, this station will be quite large—in fact it will be Arizona's largest electrical generating station. It will be the third power generating station to be built under the Western Energy Supply and Transmission Associates (WEST) concept, in which participating utilities cooperate in extensive regional planning of generation and transmission facilities and coordinate their investment in such facilities. Generating plants, much larger than any single utility would need, are constructed and operated by groups of utilities achieving economies that the participants would not otherwise enjoy. This practice helps to keep consumers' power costs low and makes protection of the environment more feasible.

The project is being managed by the Salt River Project of Phoenix, Arizona. The coal mines are operated by the Peabody Coal Company. The railroad that will be built to carry the coal to the station is being engineered

TABLE 17.1

NAVAJO PROJECT PARTICIPANTS

	PERCENTAGE OF PARTICIPATION	KILOWATTS TO BE RECEIVED (APPROXIMATE)
SALT RIVER PROJECT	21.7	501,000
LOS ANGELES DEPARTMENT OF WATER AND POWER	21.2	490,000
ARIZONA PUBLIC SERVICE CO.	14.0	323,000
NEVADA POWER CO.	11.3	261,000
TUCSON GAS & ELECTRIC CO.	7.5	173,000
U.S. BUREAU OF RECLAMATION	24.3	561,000
TOTAL	100.0	2,309,000

and constructed by Morrison-Knudsen Company, Inc. Bechtel Corporation is the engineer-constructor of the electric generating facilities.

Based upon the load growth projected by the Navajo participants, power capacity of the station is scheduled to become available in three 750 Mw. increments in the years 1974, 1975, and 1976. Construction began in April 1970.

The combination of this large size and the current timing has served to make the Navajo Project a strong focal point from the standpoint of environmental impact. The participants, well aware that this will be a test of the utility industry's ability to build large fossil-fuel power plants without causing an unacceptable degradation of the environment, have pledged themselves to meet this challenge of environmental acceptability.

SITE DESCRIPTION

Site selection criteria included the need for competitive fuel, reliable cooling water supply, nearby townsite, relationship to load, and acceptability from the standpoint of social and environmental impact.

To locate the station, let us look at several maps. First, in figure 17.1, a map of northern Arizona and environs. Next, in figure 17.2, a map showing locations of the Black Mesa coal mine, the site of the generating station near Lake Powell, and the route of the Black Mesa and Lake Powell Railroad that will be built to transport the coal from the mine to the station. We see here the relationship of the site to the Grand Canyon, Marble Canyon, and Glen Canyon recreation areas, and its position with regard to the Navajo and Hopi Indian Reservations.

A more enlarged map, fig. 17.3, shows the relationship of the site to Page, Arizona, the Glen Canyon Dam, and Lake Powell, which is the source of cooling water. Note also, the lake pumping (intake) station and the ash disposal site.

The region of the generating station site and its environs is relatively uninhabited and starkly beautiful. As is typical in this type of desert, the colors and patterns change radically at different times of day. Visibility is generally quite good—thirty miles or more.

ENGINEERING PARAMETERS AND VITAL STATISTICS

Peabody Coal Co. has coal mining leases on an area located on the Black Mesa, the Hopi Reservation (see fig. 17.2).

The total leased area covers 64,000 acres on both Navajo and Hopi land. Coal on approximately 14,000 acres will be mined at the rate of 400 acres

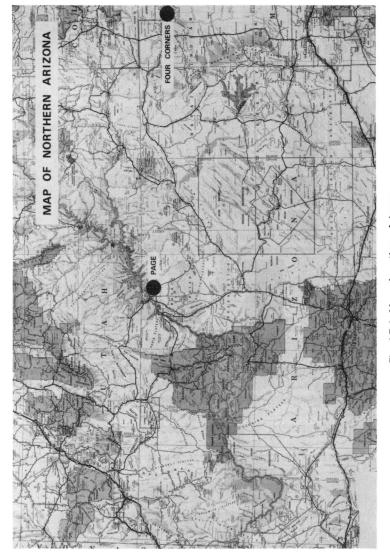

Fig. 17.1 Map of northern Arizona.

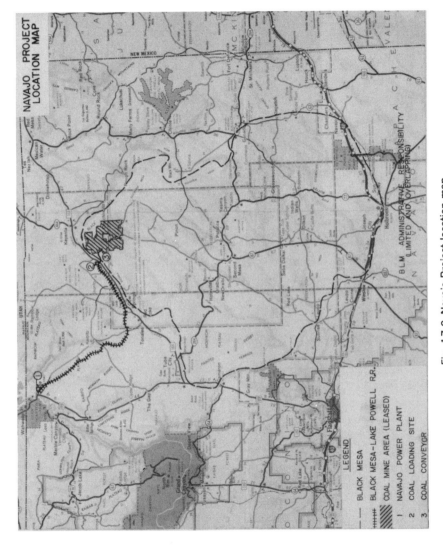

Fig. 17.2 Navajo Project location map.

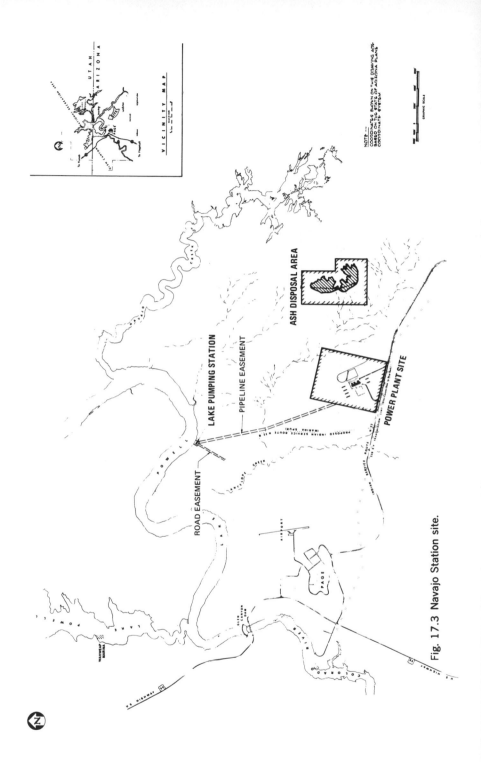

Fig. 17.3 Navajo Station site.

290

per year to supply fuel for both the Navajo Generating Station and the Mohave Generating Station (fig. 17.4). Coal for the Mohave Station is slurried and piped to the station.

The mines are operated in stripping fashion, and special efforts are planned for reclamation and reseeding of the affected areas.

The coal found in this area is a typical western sub-bituminous grade. Table 17.2 shows some characteristics of this coal. Note in particular the low sulfur fraction. The details of composition of the ash are shown in table 17.3.

The route of the Black Mesa and Lake Powell Railroad, shown in figure 17.2, will take the coal about eighty-five miles from mine to station via electric trains energized by a 50 kV catenary. The schedule calls for three round trips per day of seventy-three–car trains.

Three 500 kV transmission lines are being built. One heads west towards Nevada and Los Angeles, while the other two carry power southward towards Phoenix.

The generating station will comprise three 750 Mw. units fired by Combustion Engineering pulverized-fuel boilers. An artist's conception of the completed plant is shown in figure 17.5.

Up to 34,100 acre-feet per year (21,000 gpm.) will be withdrawn as makeup water to the closed-cycle system. Water will be pumped from Lake Powell through the condensers, cooled by evaporation in a set of mechanical draft cooling towers, and recycled. Rows of cooling towers flank the generating units.

We can also see the locations of the coal storage piles, the largest of which will have a thirty-day supply of coal. It is estimated that 24,000 tons per day will be burned, at full-load conditions. The ash will be trucked to a box canyon about two miles from the station, where it will be dumped, compacted, and stablized.

ENVIRONMENTAL REQUIREMENTS AND GOALS

Legislation of environmental impact is a rapidly expanding arena. Of course, there are regulations on almost all aspects of industrial output, such as noise, dust, etc. But the most significant, with respect to power generating stations, concerns air pollution.

Table 17.4 is a list of some of the ambient air quality standards in force in 1971 or proposed in Arizona and Utah; proposed federal ambient air quality standards are shown for comparison.

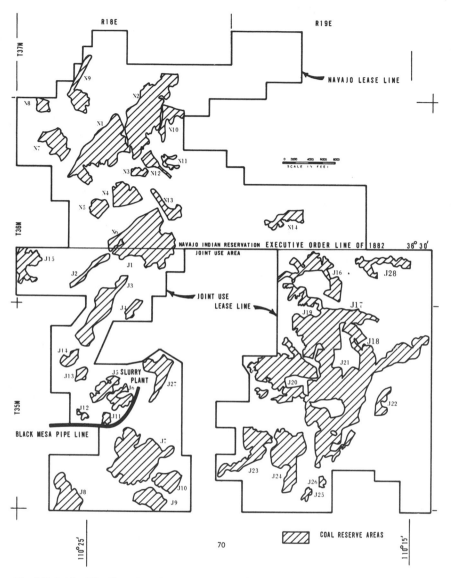

Fig. 17.4 Identification of coal reserve areas within the Peabody Coal Company leases, Black Mesa, Arizona.

TABLE 17.2

COAL CHARACTERISTICS BLACK MESA SUB-BITUMINOUS COAL

	PERFORMANCE GRADE	AVERAGE	RANGE	REJECTABLE LIMITS PER COAL CONTRACT
HEATING VALUE AS FIRED IN BTU/LB	10,725	11,000	10,000 – 11,500	10,000
ASH FUSION TEMPERATURE IN DEGREES F (RED)				
INITIAL	2235	2235	2090 – 2360	
SOFTENING	2330	2330	2205 – 2460	H=W, 2150
FLUID	2470	2470	2360 – 2650	
GRINDABILITY, HARDGROVE NO.	42	42	39.7 – 44.4	40
ULTIMATE ANALYSES IN PERCENT/WT				
MOISTURE*	10.27	10.74	9.73 – 16.0	16.0
SULFUR	0.50	0.51	0.36 – 1.26	1.34
ASH	10.43	7.93	5.84 – 12.18	12.9
CHLORINE	0.01	0.01	0.01 – 0.02	
CARBON	61.29	62.86	54.87 – 65.42	
HYDROGEN	4.37	4.48	3.91 – 4.62	
OXYGEN	12.13	12.44	10.85 – 12.95	
NITROGEN	1.00	1.03	0.90 – 1.07	
TOTAL PERCENT	100.00	100.00		
PROXIMATE ANALYSIS IN PERCENT/WT				
MOISTURE*	10.27	10.74	9.73 – 16.0	30
VOLATILE	37.94	38.91	37.41 – 40.67	
FIXED CARBON	41.36	42.42	37.85 – 43.71	
ASH	10.43	7.93	5.84 – 12.18	
TOTAL PERCENT	100.00	100.00		
ALKALI AS NaO_2, IN PERCENT				0.31

TABLE 17.3

ASH CHARACTERISTICS BLACK MESA SUB-BITUMINOUS COAL

ASH ANALYSIS	AVERAGE	RANGE
PHOSPHOROUS PENTOXIDE, P_2O_5	0.27	0.08 – 0.57
SILICA, SiO_2	52.99	47.62 – 58.91
FERRIC OXIDE, Fe_2O_3	5.77	4.41 – 9.74
ALUMINA, Al_2O_3	21.17	16.72 – 24.87
TITANIA, TiO_2	1.05	0.86 – 1.34
LIME, CaO	7.78	4.94 – 12.06
MAGNESIA, MgO	1.88	1.19 – 2.61
SULFUR TRIOXIDE, SO_3	6.34	2.16 – 10.8
POTASSIUM OXIDE, K_2O_3	0.62	0.39 – 1.19
SODIUM OXIDE, Na_2O	1.53	0.53 – 2.47
UNDETERMINED	BALANCE	

Fig. 17.5 Artist's conception of the completed power plant, Navajo Project.

TABLE 17.4

SOME AMBIENT AIR QUALITY STANDARDS
(All units are $\mu g/m^3$.)

ARIZONA

PARTICULATE
ANNUAL: 70
1 DAY: 100

SO_2
ANNUAL: 50
3 DAYS: 120
1 DAY: 250
1 HOUR: 850

UTAH

PARTICULATE
ANNUAL: 90

SO_2
ANNUAL: 53
1 DAY: 264

FEDERAL ENVIRONMENTAL PROTECTION AGENCY

PARTICULATE
PRIMARY
ANNUAL: 75
DAILY: 260
SECONDARY
ANNUAL: 60
DAILY: 150

SO_2
PRIMARY
ANNUAL: 80
DAILY: 356
SECONDARY
ANNUAL: 60
DAILY: 260

CO
8 HOURS: 10,000
1 HOUR: 15,000

OXIDANT
1 HOUR: 125

HC
3 HOURS: 125

NO_X
ANNUAL: 100
DAILY: 250

Emission restrictions are not yet fully developed. The particulate emission limits for Arizona are

$$Y = 0.6(X < 10)$$
$$Y = 1.02X^{-0.231} \quad (10 < X < 40000)$$
$$Y = 17X^{-0.568} \quad (X > 4000)$$

where

Y = allowable emissions in lbs./10^6 Btu.
X = plant capacity rating in 10^6 Btu./hr.

Arizona's definition of particulate includes vapor, other than water, that would be liquid at normal temperature and pressure.

In addition to legislated restrictions, the Navajo participants are obligated by various contractual agreements to maintain both air and water quality of an even more stringent degree. The contract with the United States Department of Interior allowing the use of the water gives the department approval authority over waste water and waste materials disposal. In addition, it restricts water return to the lake to no more than 5,900 acre-feet per year at a temperature no greater than 90° F. Particulate removal equipment for flue-gas must operate at 97 percent removal efficiency or better. Coal must have less than 1.5 percent sulfur and less than 14.5 percent ash.

The lease of land from the Navajo Tribe adds to these requirements that the particulate removal equipment must be designed for 99.5 percent removal efficiency. This requirement is reiterated in the agreement among the six participants of the Navajo Project.

Today there is much more to environmental control than merely meeting legislative or contractual standards. The Navajo Project will be the first large coal-fired station to be constructed in the Southwest since the restoration and enhancement of our environment became a national goal. The participants are highly sensitive to this and are striving to meet this challenge. They have already decided not to return any water to the lake. They have established a policy that SO_2 removal equipment will be installed as soon as it is commercially available. They have even initiated a research program in SO_2 removal to accelerate the development of commercial availability. A study program is being formulated to investigate effects of emissions upon visibility in the area. A program to determine the long-term effects upon the ecology of the site environs is also under development.

APPLICATIONS OF ENVIRONMENTAL IMPACT-ABATEMENT TECHNOLOGY

The Colorado River is the major source of water for a large area of North America. We have recently begun to feel the weight of abusive use of its resources. Therefore, at the outset, once-through cooling was considered an intolerable environmental (thermal) burden upon the river. The water contract thus permitted up to 40,000 acre-feet/year to be withdrawn, with a maximum of 5,900 acre-feet/year to be returned.

An intensive study was initiated to predict what the effects on Lake Powell might be of the intake and subsequent cooling-tower blowdown return. We proceeded along two fronts. The hydraulics and thermodynamics of intake and return were calculated, using measured data on existing lake parameters in computer programs of plume dispersion. At the same time, ecological studies of the existing lake inhabitants, as well as predictions of the future of the lake, were made.

Figures 17.6 through 17.11 show examples of the results of the hydraulic computer programs. These show in two dimensions the depth and distance to which the water of the lake might deviate from the ambient values of total dissolved solids and temperature, at different times of the year. For these causes, it was assumed that the blowdown water was concentrated in the cooling towers by a factor of six, before discharge to the lake.

We see that the maximum plume centerline dissolved solids concentrations in the blowdown are reduced by approximately 80 percent within fifty feet from the orifice. The temperature reduction depends upon ambient air wet-bulb temperatures as well as local ambient temperature, hence the difference between summer and winter curves.

From this type of calculation we have been able to conclude that the *detectable* plume would be confined to an area well under 15 percent (300 feet) of the reservoir cross section and within 200 feet upstream and 500 feet downstream of the outlet point. Even within this volume, high concentrations and temperatures would prevail only within twenty-five feet of the outlet point.

Estimates of the effect on total dissolved solids downstream of the plant indicated that maximum increases of less than 3 ppm. (from 650 ppm. to 653 ppm.) could occur.

The ecological study of the lake indicated that, since the impoundment of the river, the fish fauna have been and are still in a constant state of change as new kinds become established and begin reproducing. Population stability is not expected to be achieved for a decade or longer. It was possible to say, however, that due to the relatively small volumes of return and the quick entrainment and dilution, none but the most local effects upon the aquatic biota of the lake could be expected.

However, in order to avoid establishing a precedent which, if followed by other water users, would be detrimental to the users of Lower Basin waters of the Colorado River, the Navajo participants elected not to return any water to the lake.

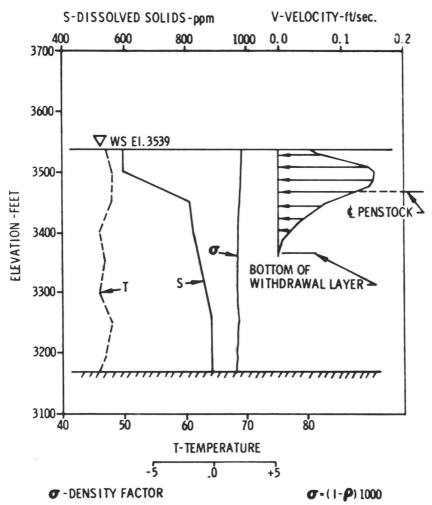

Fig. 17.6 Typical plume effect in winter lake characteristics.

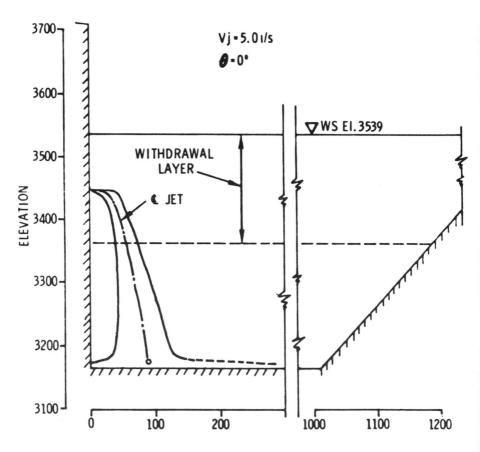

LAKE POWELL

Jan 13, 1969

$V_j = 5.0 \, l/s$

$\theta = 0°$

∇ WS El. 3539

WITHDRAWAL LAYER

₵ JET

ELEVATION

3700
3600
3500
3400
3300
3200
3100

0   100   200   1000   1100   1200

RESERVOIR CROSS SECTION AT NAVAJO PLANT

Fig. 17.7 Typical plume effect in winter plume dispersion.

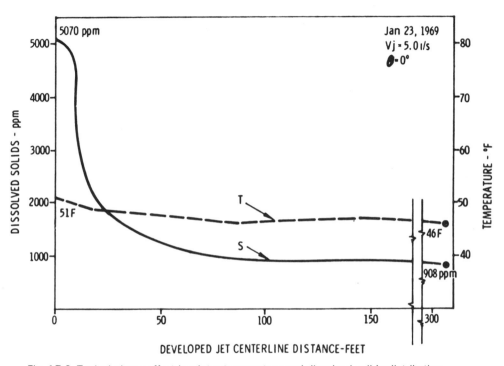

Fig. 17.8 Typical plume effect in winter temperature and dissolved solids distribution.

301

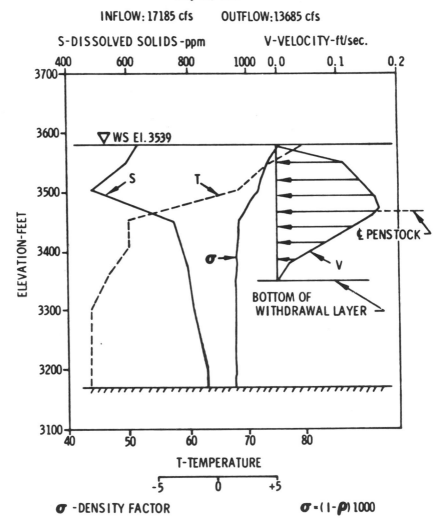

Fig. 17.9 Typical plume effect in summer lake characteristics.

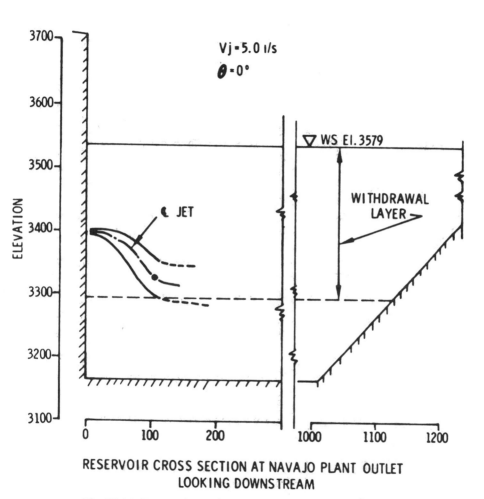

Fig. 17.10 Typical plume effect in summer plume dispersion.

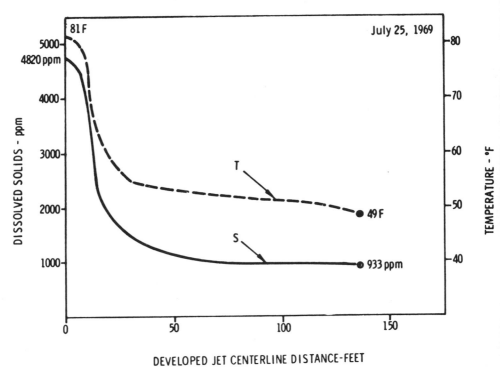

Fig. 17.11 Typical plume effect in summer temperature and dissolved solids distribution.

The question of how to handle the cooling tower blowdown, which may amount to 3300 gpm., has not yet been decided. The alternatives include:

1. Large evaporation ponds
2. Purification by distillation plus a small pond
3. Purification by reverse osmosis plus a small pond
4. Purification by electrodialyses plus a small pond
5. Utilization in wet scrubber particulate remover

The first four alternatives are being evaluated on an economic basis. The last alternative would add no cost at all, but depends upon whether or not wet scrubbers will be used.

The essential fact remaining is that present technology permits the complete protection from degradation of water quality. The costs involved are in the form of both water consumption and purification equipment.

The greatest emphasis on pollution abatement in fossil-fueled power plants is being placed upon air quality control. The magnitude of this task can be seen by referring to the table of emissions shown in table 17.5. Considerable amounts of material in the form of particulates, SO_2 and NO_x, are emitted from the boiler. These are all undesirable in high atmospheric concentrations, hence it is important to absorb as much as possible and disperse the remainder as efficiently as possible. This, incidentally, is our answer to the oft-debated argument comparing high stacks to removal equipment.

For particulate removal we are purchasing equipment designed to remove 99.5 percent by weight. Special redundancy and sectioned design will be built in so that any single section may be repaired during full-power operation, during which the efficiency will not fall below 99 percent. Currently under consideration for this are electrostatic precipitators and wet scrubbers.

Electrostatic precipitators have the advantage of many years of power generation application. There is reason to believe that, with proper design and operational maintenance, these devices will do the job. However, there has been very little experience in meeting the high removal efficiency required here for long periods in a large coal-fired application. In addition, the very low sulfur content (0.5 percent average with range from 0.3 percent to 0.8 percent) raises questions concerning the proper approach to take. The high electrical resistivity of this coal tends to decrease the overall collection efficiency of a precipitator, mainly by permitting the buildup of charge on the plates, reducing the collecting field.

There are three possible methods of handling this aspect of low sulfur ash collection:

TABLE 17.5

ESTIMATED STACK EMISSIONS - NAVAJO GENERATING STATION

(POWER LEVEL 2300 MW)

	TONS/DAY	PPM (VOL)
	AVERAGE	AVERAGE
COAL		
COAL CONSUMED	24,000	—
AIR CONSUMED	227,000	—
FLUE GAS (7.2 X 10^6 CUBIC FEET/MINUTE, AT 250°F AND 0.85 ATMOSPHERE)		1,000,000
CO_2	52,700	142,000
H_2O	14,600	96,500
O_2	7,850	29,200
N_2	172,000	731,000
SO_2	210	390
SO_3	1	2
NO_x (AS NO_2)	204	550
HC (AS METHANE)	1	7
FLY ASH:		
0% REMOVAL	1,450	
99% REMOVAL	14.5	
99.5% REMOVAL	7.25	

1. Increase the precipitator size
2. Inject SO_3 gas (or other conductive gas) upstream
3. Design for operation on the hot side of the air preheater

The first technique has been successful in Australia and is the preferred method there. The second approach, which supposedly coats the ash with a conductive film, was pioneered in England, and has successfully increased precipitator efficiencies both in England and the United States. The third method takes advantage of the natural decrease of resistance with increased temperature. Each of these has advantages and disadvantages of operation. At present no clear-cut superiority of one approach over the others has been demonstrated.

Another type of particulate remover, the filter bag (fig. 17.12), is also capable in principle of achieving high removal efficiencies. At present, as there is not yet any full-scale experience on a coal-fired station, the filter bags are not being considered for this application.

A relatively new competitor, at least for the power industry, is the wet scrubber (fig. 17.13). These are inherently more expensive than precipitators and require significant operating power to push the flue gas through them. While there is much experience in industries that do not have the large flue gas volumes and fine particle size of the coal-fired power plant, there has not yet been a full-size demonstration of high efficiency removal on the latter. Currently, there is considerable development under way, but as yet it is extremely difficult to predict the design requirements, economics, and control and maintenance problems of wet scrubbers requiring 99.5 percent efficiency. An important reason for the development of these scrubbers is that they may be able to double as SO_2 absorbers.

It is well known that the sulfur content in fossil fuels leads to the emission of SO_2 which, in high concentrations, can be detrimental to visibility, agriculture, and health. Much effort is being expended currently on finding low-sulfur oil, meaning oil having a sulfur content of less than 1 percent. Where low sulfur fuel cannot be found, very tall stacks are used to disperse the SO_2 to low concentrations before reaching ground levels. In addition, we see a rapid growth of the technology of removing SO_2 from flue-gases when burning high-sulfur fuels (3%–5% S).

The diversity of the state-of-the-art of SO_2 removal technology bears close resemblance to that of nuclear reactor technology in the early 1950s. There are several dozen processes being proposed and tested by various companies. None have yet been demonstrated on a large scale. The methods being tested vary in sophistication from relatively crude to highly exotic. The crude method involves the use of an alkaline scrubbing liquor to absorb the SO_2. A schematic diagram of this is shown in Figure 17.14. In the scrubber,

Fig. 17.12 Typical filter bag dust collector.

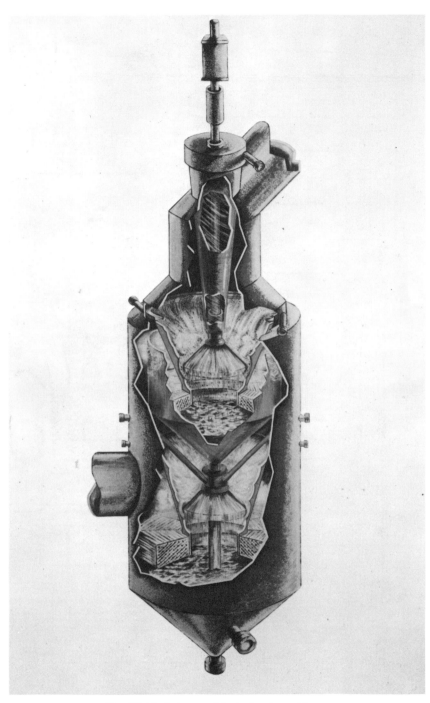

Fig. 17.13 Cutaway of typical wet scrubber.

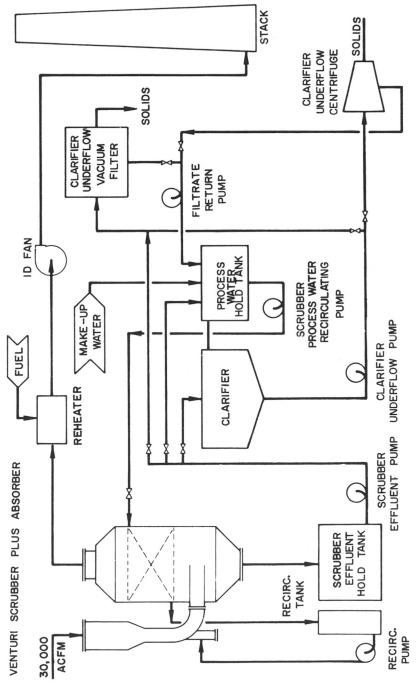

Fig. 17.14 Sulfur removal process using alkaline scrubbing.

310

the flue gas is forced to mix intimately with the aqueous liquor (either a slurry or a solution), permitting transfer of both fly-ash and SO_2 from the gas phase to the liquid phase. The slurry of fly-ash, unused alkali, and the sulfates must then be disposed of as a solid waste.

The more exotic variations, some using liquid metal absorbents or solid catalysts, are designed to yield a marketable product such as sulfuric acid, ammonium sulfate, liquefied SO_2, or elemental sulfur.

It is impossible at the moment to predict which of these processes will predominate in the long run. However, the alkaline scrubbing system is the most highly developed, and its associated costs do not depend upon the vagaries of the by-product market. This is not to say that there are no significant problems remaining to be solved. Existing demonstration units at Union Electric Company and Kansas Power and Light have been plagued with a variety of difficulties, including corrosion and plugging of the internals.

With respect to application at the Navajo Station, it should be noted that we already have a low-sulfur fuel. The coal has a sulfur content ranging from 0.3 percent to 0.8 percent, with an average value of 0.5 percent. Some of this sulfur is in the form of pyrites, which are removed to a degree in the pulverizing process. As we will show later, very tall stacks are being provided to disperse the flue gases. Nevertheless, installation of SO_2 flue-gas removal equipment is currently planned for the station. Thus all three types of SO_2 concentration reduction techniques will be applied.

Due to the relatively undeveloped status of SO_2 technology, especially as applied to low-sulfur applications and the desire of the participants to accelerate development for use at the Navajo Station, they have embarked on a program of experimental research and testing, along with the Mohave Generating Station participants (which adds the Southern California Edison Company to the list of participants mentioned earlier). The program involves the testing on a pilot scale of four types of scrubbers and three different variations of alkaline absorption processes. A highly instrumented test facility will be constructed at the Mohave Generating Station at Bullhead City, Nevada. This station burns the same Black Mesa coal as will be used at the Navajo Station. The processes tested will include the limestone slurry, lime slurry, and soda-ash solution systems. The process flow diagram using lime or limestone is shown in figure 17.15. In this series of experiments, we are interested in determining the removal efficiencies for this low SO_2 concentration. These have been shown, for high-sulfur applications, to range from 80 percent to 90 percent. However, as we see in figure 17.16, the removal efficiency will vary for different inlet sulfur conditions. If one were to assume a constant outlet concentration, the removal efficiency would decrease quite strongly. On the other hand, we know that this assumption is not true, nor is the assumption of constant efficiency. The

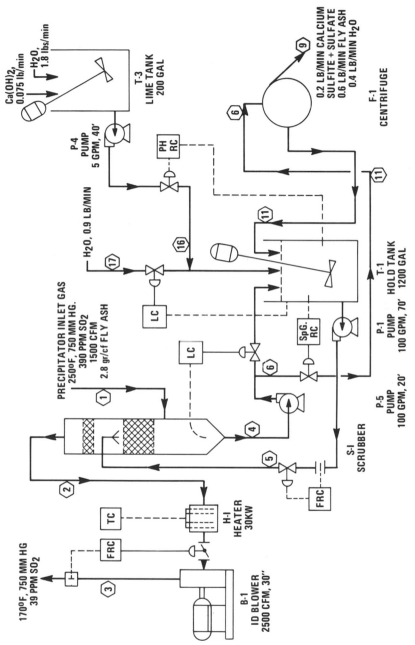

Fig. 17.15 System I—calcium absorption.

Ca(OH)₂, 0.075 lb/min
H₂O, 1.8 lbs/min

T-3 LIME TANK 200 GAL

P-4 PUMP 5 GPM, 40'

PH RC

0.2 LB/MIN CALCIUM SULFITE + SULFATE
0.6 LB/MIN FLY ASH
0.4 LB/MIN H₂O

F-1 CENTRIFUGE

H₂O, 0.9 LB/MIN

SpG. RC

T-1 HOLD TANK 1200 GAL

P-1 PUMP 100 GPM, 70'

P-5 PUMP 100 GPM, 20'

PRECIPITATOR INLET GAS
2500°F, 750 MM HG.
390 PPM SO₂
1500 CFM
2.8 gr/cf FLY ASH

S-1 SCRUBBER

H-1 HEATER 30KW

B-1 ID BLOWER 2500 CFM, 30"

170°F, 750 MM HG
39 PPM SO₂

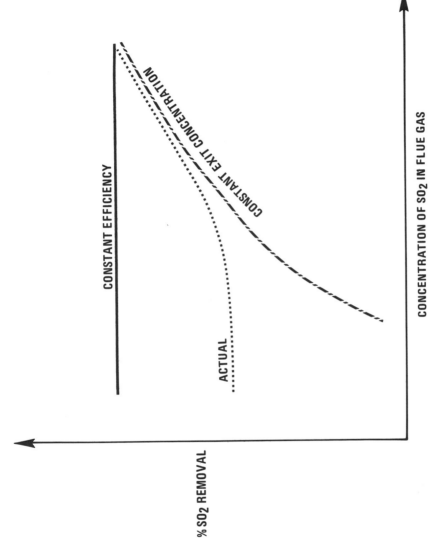

Fig. 17.16 Effect of inlet sulfur concentration on removal efficiency for a given wet scrubber with fixed operating parameters.

313

true case, which is too difficult to calculate due to the large number of nonequilibrium chemical reactions occurring, is represented by the dotted curve, which approaches the constant outlet concentration at high values and the constant efficiency at low values of inlet concentration.

The third process being tested, the soda-ash solution, is shown schematically in figure 17.17. It is well established that this liquor will absorb SO_2 with higher efficiency than the slurries mainly due to the high solubility in the aqueous solution. It is necessary, however, for economic and environmental reasons, to regenerate the sodium solution rather than discard it. Therefore, there is an extra loop in this process in which we shall attempt to promote reactions between the sodium sulfites and lime to produce calcium sulfate, which in turn will be discarded. The sodium solution will then be returned to the absorber.

It has not yet been established under what conditions the regenerative reactions alluded to above will occur. This will be determined in the test loops.

At the same time, we shall try to determine the effects of these wet scrubbing processes on the removal efficiencies of such trace elements as fluorine and mercury. There is increasing concern about the presence of these elements in the atmosphere. We are currently measuring the extent of trace elements in the Black Mesa coal to determine whether the Navajo Station will emit these in significant amounts. First reports of measurements of mercury indicate only 0.02 ppm. in the coal. We have already measured the uranium content of the ash, finding amounts similar to those measured elsewhere (table 17.6). These levels would result in maximum radioactivity concentrations in air calculated to be about 0.2 percent of the maximum permissible values established by the AEC.

Next are the oxides of nitrogen, or NO_x emissions. There has been a considerable development of the use of combustion control to reduce the formation of oxides of nitrogen in boilers. These techniques, pioneered by Southern California Edison Company, include staged combustion as well as recirculation of flue gases. These are highly successful with gas-fired boilers and somewhat less so with oil-firing. Results with coal-fired boilers have been contradictory and not successful in general. The differences involve the type of combustion (solid vs. gaseous), the internal nitrogen content of both oil and coal, and finally, the methods of power level control in the various types of units. The boiler manufacturer, Combustion Engineering, is working quite actively in attempting to reduce the NO_x formation. Although no methods have yet been proven, over-fire air ports are being installed in the boilers in order to permit their use as secondary air inputs, should this approach to staged combustion be feasible.

There have been reports that SO_2 scrubbers remove some of the nitrogen

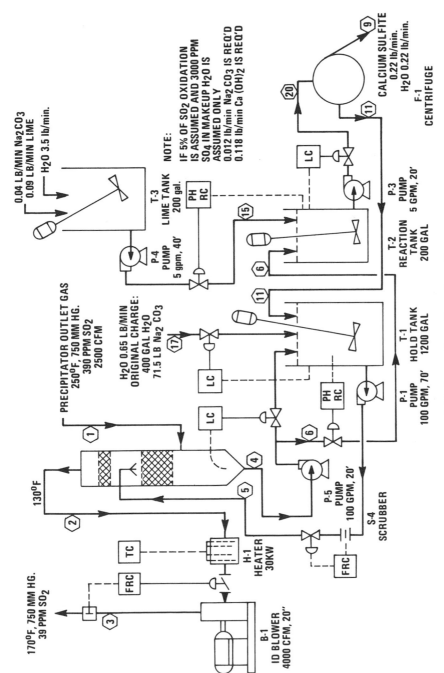

Fig. 17.17 System II—sodium absorption.

315

TABLE 17.6

MAJOR RADIOACTIVE ELEMENTS IN BLACK MESA ASH

Ra^{226}	2.6	pCi/gm
Ra^{228}	4.7	pCi/gm
Th^{228}	4.7	pCi/gm
Th^{232}	4.7	pCi/gm
U^{238}	2.6	pCi/gm

NOTE: pCi = 10^{-12} CURIES

oxides. This prospect will be examined as part of the research program previously described.

Yet another aspect of air quality with which we are concerned involves visibility. This is, oddly enough, one of the more complex problems to analyze. While there are no regulations concerning visibility, other than directly at the stack effluent, this is felt to be one of the effects considered most important by the general public. Great concern has been expressed about the possible effects of the Navajo Station upon visibility in the Grand Canyon and other recreation areas. We all know what visibility is, intuitively, and yet establishing quantitatively the meaning of visibility is a task on the forefront of atmospheric sciences. It is not surprising then that predicting the effect upon visibility of the discharge of a plume into the atmosphere is even more difficult. But even such predictions are based mostly upon evaluation of the effects of the materials as emitted from the stack. This may be termed the primary effect. However, it is known that chemical reactions occurring after emission can be more important. The analysis and measurement of this secondary effect is also an objective of current research in this field. The Navajo Project is currently formulating a research program as well, in order to accelerate these developments, predict as accurately as possible the effect of the station, and apply every means available to reduce the impact.

The Navajo participants' answer to the question—tall stacks or abatement equipment—is both. The method for determining stack height involved the following steps:

1. Analysis of expected emissions
2. Measurement and analysis of meteorological dispersion data, including atmospheric stability, wind structure, and ambient temperature
3. Development of design criteria
4. Calculations of dispersion as a function of stack height

Of course, steps 2 through 4 actually followed in an iterative manner. It is not necessary to dwell on the relatively straightforward aspects of steps 1 and 2. It turns out that with high-efficiency particulate removal and the present ambient air quality standards, SO_2 dispersion is the critical factor even with low-sulfur fuel. This has been observed before and is probably true in general.

It appears impossible to show that the present SO_2 ambient standards can be met under all atmospheric conditions by use of tall stacks alone. Observe that meteorologists have been able to classify three general types of atmospheric models pertinent to plume dispersion: fumigation, limited mixing, and neutral.

Fumigation, sometimes called "inversion-breakup," occurs often in the mornings after a plume has been maintained under an inversion layer for the preceding hours (fig. 17.18). The process of breaking the inversion begins with heating from the ground up. When the heated and turbulent air reaches the plume for a brief period, usually less than thirty minutes, the plume may be locally brought to the ground level in high concentrations. These intensities are relatively insensitive to variation in stack height, unless such variation determines whether the plume is above or below the nocturnal inversion height.

The second condition, termed limited mixing, would occur during relatively rare periods in which air circulates quite well under an inversion which acts as a "roof." As illustrated in figure 17.18, the plume tends to disperse relatively uniformly between the "mixing level" and the ground. This can occur over a wide area and can last for periods of several days. Once again, the stack height has little or no effect upon the concentrations that may be reached during such a limited mixing condition.

In the neutral model (fig. 17.18), the plume rises due to its bouyancy until it reaches some equilibrium level, which depends upon the wind speed, and then continues to move in an ever-expanding and diluted plume. In this model, as we shall see, the stack height affects the maximum ground concentrations quite strongly.

All of the foregoing indicates why the stack height selection criteria were based on neutral dispersion conditions only. It is implied that, to the extent that the concentrations during the other two conditions are intolerable, they must be reduced by means other than stack height increase. Thus, the design criterion was that the SO_2 ground level concentration should never exceed the permissible Arizona limits under neutral dispersion conditions.

A digital computer program was used to calculate the combined plume rise and subsequent dispersion. Concentric circles and radii with the plant site as the center were used to divide the area into 144 sectors, each having its own elevation based upon topography. The maximum concentrations were calculated at each sector for all wind velocities up to twenty-five meters per second. Part of a typical computer output is shown in table 17.7 for two stack heights. This is repeated for various stack heights, so that a plot of maximum concentrations as a function of stack height could be made (fig. 17.19). Application of the design criteria to this curve, which involved converting the one-hour standard of 850 $\mu g/m^3$ at S.T.P. to the altitude and mean temperature at the site, which yielded 695 $\mu g/m^3$, led to the adoption of 775 feet as the stack height.

A large array of monitoring stations has been set up around the plant site to obtain ambient background data before plant operation. Locations have been determined on the basis of a desire to check the calculations of meteorological dispersion at key points, as well as to document levels of

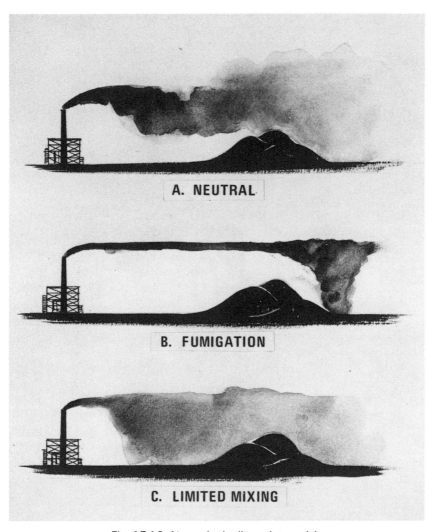

A. NEUTRAL

B. FUMIGATION

C. LIMITED MIXING

Fig. 17.18 Atmospheric dispersion model.

TABLE 17.7

PORTION OF TYPICAL COMPUTER OUTPUT FOR SULFUR CONCENTRATIONS

STACK HEIGHT = 700 FEET

DIRECTION		DISTANCE IN KILOMETERS					
		1	3	5	7	10	20
SE	HRS	0	0	0	0	0	0
	XMX	0	0	87	212	418	267
	SMX	25	25	25	25	25	22.7!
SSE	HRS	0	0	0	12	0	0
	XMX	0	1	131	957	739	398
	SMX	25	25	25	22.75	20.5	18.75
S	HRS	0	0	0	0	0	0
	XMX	0	2	192	443	550	386
	SMX	25	25	25	25	25	20.25

STACK HEIGHT = 800 FEET

DIRECTION		1	3	5	7	10	20
SE	HRS	0	0	0	0	0	0
	XMX	0	0	36	119	303	226
	SMX	25	25	25	25	25	23.75
SSE	HRS	0	0	0	0	0	0
	XMX	0	0	56	810	655	363
	SMX	25	25	25	23.5	21	20.5
S	HRS	0	0	0	0	0	0
	XMX	0	0	87	276	418	348
	SMX	25	25	25	25	25	20.75

HRS = NUMBER OF HOURS PER YEAR CONCENTRATION EXCEEDS 850

XMX = MAXIMUM SO_2 GROUND LEVEL CONCENTRATION IN $\mu G/M^3$

SMX = WIND SPEED CORRESPONDING TO XMX IN M/SEC

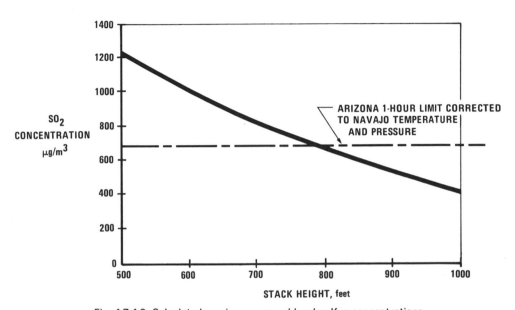

Fig. 17:19 Calculated maximum ground-level sulfur concentrations.

the important contaminants at selected scenic, historical, and recreational areas. In figure 17.20, we see the monitoring points currently set up. These are single passive monitors, as shown in figure 17.21, carrying a dustfall jar and a sulphation "candle." These give relative intensities of particulates and SO_2 as averaged over a monthly period.

In addition to these outlying stations, a more sophisticated meteorological tower, sampling station, and laboratory have been established at the site. These have active samplers for particulates, SO_2, NO_x, and oxidants (mostly ozone). Measurements thus far have provided input data for the stack plume dispersion calculations and have documented the low pollution and high visibility of the area. Somewhat unexpectedly, oxidant levels are about twice as high as other areas with low industrial activity. However, at 0.06 ppm., this is still quite a low value.

ECOLOGY

It has been postulated that a power plant of this size would have an impact upon the ecology of the surrounding area. It is known that high concentrations of pollutants can kill plants, poison soil, or simply induce animals to leave the neighborhood. The question thus becomes: What is the significance of low levels of pollutants on a long-term basis? To answer this question, one must make measurements in the field. However, there is typically complex change and it is difficult to relate the effects to the causes. There is some hope that in this desert area the effects, if any, of a large power plant may be discernible, especially if an ecological monitoring program is established well before operations are begun. Such is the nature of the ecological study program proposed by a joint team from Brigham Young University and Northern Arizona University. Observation would be made on a dozen test "quads" arrayed around the plant site over a period beginning three years before initial operation. Measurements of microscopic, as well as macroscopic, aspects of the biosphere would be made, along with careful recording of the climatological and chemical aspects of the environment. The Navajo Project participants recognize the desirability of such a study and are currently evaluating details of the proposal.

NOISE CONTROL, DUST CONTROL, ASH DISPOSAL, AND AESTHETICS

The remaining tasks that are part of the environmental engineer's responsibility should not be considered minor, although they are less dramatic, and, perhaps, more conventional in relation to the previously described

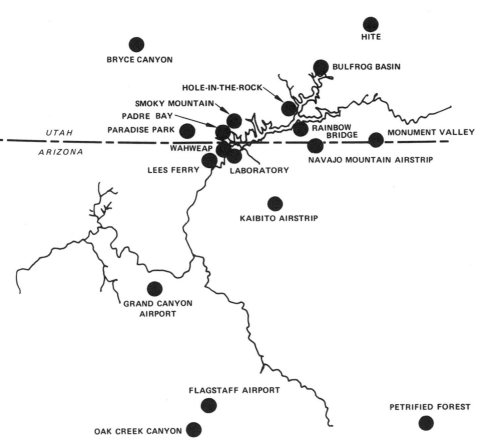

HITE

BRYCE CANYON

BULFROG BASIN

HOLE-IN-THE-ROCK

SMOKY MOUNTAIN

PADRE BAY

PARADISE PARK

UTAH

ARIZONA

WAHWEAP

RAINBOW
BRIDGE

MONUMENT VALLEY

LEES FERRY

LABORATORY

NAVAJO MOUNTAIN AIRSTRIP

KAIBITO AIRSTRIP

GRAND CANYON
AIRPORT

FLAGSTAFF AIRPORT

PETRIFIED FOREST

OAK CREEK CANYON

Fig. 17.20. Map of monitoring station locations.

Fig. 17.21 Single passive monitors at stations around plant site.

aspects. These tasks consist of noise control, dust control, ash disposal, and aesthetics.

The plant layout has taken into account the self-absorption of some of the sound that might otherwise have been transmitted toward Page. Records of ambient sound levels will be taken for documentation and analysis prior to plant operation. During design, predictions of generated sound levels will be made and designs will attempt to minimize them. Then, after completion, surveys will be taken again to show where further abatement techniques may be required.

Coal generating stations can have severe dust problems. Both coal and ash movement tend toward dust releases. Wherever feasible, it is planned to use covered transport systems. At transition or switching points where dust could be produced, special watering equipment will be added.

The ash will be transported to an ash disposal area about two miles east of the plant (fig. 17.3). This area is a box canyon having sufficient volume to contain the ash resulting from thirty-five years of plant operation. The ash will be deposited in alternate layers at the disposal area with an earth covering. The ash and earth layers will be compacted to a high density to stablize the mass, preventing blowaway during the periods of high winds that sometimes occur in the area.

Check dams and intercepting trenches will be constructed at the downstream toe of the ash-disposal area embankments to intercept and pond any water that might run off the ash as a result of rain, snow, or moisture in the ash itself.

Aesthetics are considered to be an important aspect of environmental acceptability. Great care will be taken to ensure that the presence of the Lake Pumping Station does not intrude upon the lake itself. Therefore, although it would be most economical to cantilever the intake pipes over the edge of the lake, and it would even be feasible to cover them, the participants have elected to slant-drill through the rock for the intake pipes (fig. 17.22) and to put the pumps in an excavated area on the ground so that that part of the plant will not be visible from the lake.

As can be seen in the artist's rendering of the power station itself (fig. 17.5), the building, while not attempting to masquerade as something other than a power plant, has clean form and lines. It is believed that the station will be considered an attraction and, therefore, a visitor's center is planned.

ENVIRONMENTAL APPROVALS

In addition to meeting both federal and state legislated requirements, the Navajo Project is obligated to submit for the approval of the Department

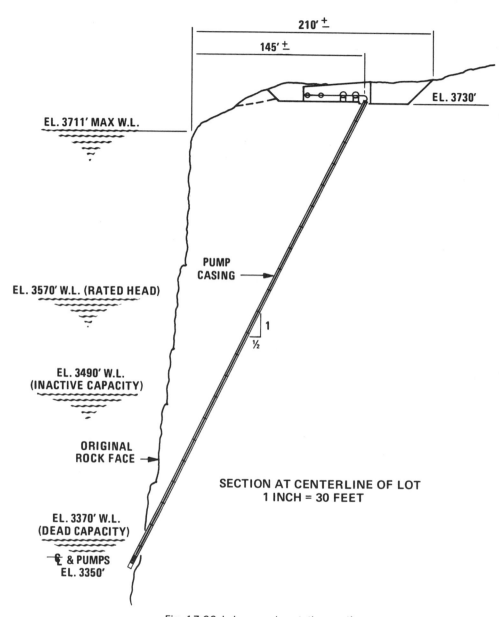

Fig. 17.22 Lake pumping station section.

of Interior, as well as the state of Arizona, all plans and designs involving environmental controls. The list of approval items is as follows:

1. Stack height
2. Lake pumping station
3. Ash handling equipment
4. Ash disposal area
5. Sanitary sewage
6. Process waste disposal
7. Particulate removal equipment
8. SO_2 removal equipment
9. Coal handling facility
10. Waste water drainage and discharge
11. Coal pile stabilization
12. Pollution monitoring instrumentation

These are required prior to construction of each item. Thus far, tentative approval has been received for the first item.

COLORADO PLATEAU ENVIRONMENTAL ADVISORY COUNCIL

As a direct result of some public meetings called to discuss the environmental aspects of the Navajo Project, a new organization was formed. This group, the CPEAC, has a membership of close to a hundred people representing academic, industrial, governmental, and private environmentalist organizations. All are interested in the preservation of an acceptable environment on the Colorado Plateau. The CPEAC, with headquarters located at the Museum of Northern Arizona in Flagstaff, has adopted as a goal the coordination of information relative to all aspects of environmental effects and the encouragement of meaningful research and action concerning the Colorado Plateau.

We have great hopes that this novel organization will serve as an important forum and focal point for establishing a flow of information between the various subgroups that make up the public.

LESSONS OF THE NAVAJO PROJECT

The Navajo Project, when completed, will be one of the world's largest electric generating stations. Because of its location and timing, it is probably one of the most ambitious projects to date in terms of environmental impact study and abatement. It will certainly be a valid test of our belief that through sincere effort and careful engineering, an electric generating station need not be a blot on the environment.

We can already extract very useful lessons from the work done thus far. One clear fact is that the breadth of environmental engineering concerns can be no less than the breadth of the impacts themselves, which cover the spectrum from A (Archaeology) to Z (Zoology).

Despite its truth, it is not sufficient to fall back upon the fact that little is known about many types of environmental impact. Wherever possible, and especially where economies of plant size are present, effort and funds should be expended to advance the knowledge in the most important areas.

Early and frequent meetings and discussions with the public are necessary to arouse public response sooner rather than later. These dialogues serve to establish clearly the most important concerns of the public.

In the case of the Navajo Project, we found that while it will be necessary to utilize the very latest techniques of pollution abatement of all types, the areas of SO_2 removal, visibility effects, and ecology require even more vigorous development efforts. Therefore, in each of these areas, programs of research are either under way or being formulated.

We estimate that the costs associated with these aspects of the Navajo Project, based solely upon environmental acceptability, amount to approximately 90 million dollars. That this huge amount can be tolerated at all is testimony to the basic economy of the concept. However, this expenditure is real, and will be borne eventually by the electricity users. We are making every effort to see that the apportionment of this money is in their interest. We sincerely believe that the benefits of electrical energy need not be accompanied by environmental degradation, that the Navajo Project will show this and will be, in addition, a source of vital feed-back information for enhancement of the environment.

18. University of California System and Air Pollution Research

Ernest S. Starkman
and
Roger B. Sperling
University of California

Perhaps a better title for this subject would have been "Anatomy of a Multi-Disciplinary Research Program" because the present title might imply that the topic will concern air pollution caused by the University of California. While this is a valid subject (the University does operate vehicles, steam plants, and other equipment and facilities which pollute the air, and steps have been taken to bring most of these into compliance with local air pollution ordinances) the topic at hand relates to the larger air pollution problem in California—often called "Los Angeles (or California) Smog" —and how the University is contributing to potential solutions.

Our purpose here will be to describe how, during the past sixteen months, the University has attempted to launch a unique statewide program in air pollution research. The subject is divided into three sections: background, current operations, and the future. Taken together, it will be shown how Project Clean Air, as this program is called, came into existence, how it has operated, and what its future seems to be.

BACKGROUND OF PROJECT CLEAN AIR

The story of Project Clean Air must begin by building a foundation of essential facts about air pollution in California and research at the University over the past two and one-half decades.

329

The Forties

The University has been identified with photochemical air pollution since the time it was "discovered." As early as 1944, plant pathologists from the University of California Citrus Experiment Station at Riverside began studying plants that appeared to have been attacked by a new disease. From studies on spinach and lettuce in the Los Angeles area they concluded that the plant damage they observed was not caused by disease but by air pollution. Later, starting in 1949, three of the Riverside scientists, Drs. Ellis Darley, James Kendrick, and John Middleton, joined in a cooperative effort with Dr. A. J. Haagen-Smit, of the California Institute of Technology in Pasadena, which resulted in the first valid explanation of the overall mechanism involved in the formation of photochemical smog. This research team, headed by Haagen-Smit, demonstrated that irradiation of hydrocarbons, oxides of nitrogen, and air produced a mixture which caused plant damage identical to that observed in the Los Angeles area.[1]

The Fifties

Air pollution research slowly grew during the 1950s in a haphazard, unorganized way. Individual faculty members on six University of California campuses conducted research projects in their laboratories. For years no effort was made to pull together research teams on one campus let alone coordinate the entire multi-campus system. In 1953 the first state funds were allocated to the University for air pollution research. The Riverside campus received 50,000 dollars for air pollution studies; this figure was increased to 150,000 dollars per year by the end of the decade and was largely used to extend the pioneering plant science and air chemistry work of UC Riverside scientists. Similarly, the College of Engineering at the Los Angeles campus was receiving special state appropriations. Dean L. M. K. Boelter was an early advocate of strong involvement of University scientists in air pollution research. This budget reached 120,000 dollars per year in 1959. A variety of engineering studies was supported from these funds, including early work on automobile engine control devices. At the end of the decade, mounting public concern resulted in University-sponsored statewide meetings of air pollution researchers and increased cooperation among groups of University scientists on their own campuses.

The Sixties

The 1960s saw an upswing in private and public monies being channeled to the study of air pollution problems. State appropriations grew modestly, allowing more faculty and staff time to be assigned to air pollution research problems. The statewide Air Pollution Research Center was organized on the Riverside campus around the core of scientists who had helped make the fundamental discoveries about smog ten years earlier. John Middleton was its first director. The amount of state funds reached a peak of 375,000 dollars per year for the two campuses in 1966 and then slowly declined. But, the state did not support even a major share of the expanding research activities of the 1960s. Money from the private sector—mainly industry and agriculture—financed a large number of research programs throughout the UC system. The major source of funding, however, was the United States Public Health Service, which was primarily responsible for a deepening and broadening of air pollution studies in California. A large portion of these research funds, administered later by the Department of Health, Education, and Welfare (through the National Air Pollution Control Administration), and now by the new Environmental Protection Agency (through the Air Pollution Control Office), came to California. In 1969, for example, one out of every six dollars spent by the United States government on air pollution research grants (not including contracts) was awarded to the University of California. One out of every eight federally financed air pollution grants was directed by a UC researcher. At the end of the decade over 100 University faculty members were engaged in basic research on air pollution, utilizing about 2 million dollars per year of state, federal, and private money in grants and contracts.

The University played an important role in California's battle against air pollution for twenty-five years. But the battle had obviously not been won. The scientific and political efforts of these twenty-five years had failed to suppress the growth of the ubiquitous blanket of smog. John Middleton, who had gone to Washington to organize the National Air Pollution Control Administration, summarized the state of air pollution control technology on motor vehicles in a Los Angeles speech on 6 December 1969:

> In my judgment, the best that we can expect from the standards now in effect is that hydrocarbon and carbon monoxide emissions will in 1980 dip to approximately 60 percent of current emissions, or roughly what they were in 1953. And after 1980, when these standards have passed the saturation point of their effectiveness, as vehicle use continues to increase, the levels of pollution will resume their upward climb.
>
> We cannot in good faith, then, report that we are winning the war with motor vehicle pollution. Rather, we are engaged in a holding action.[2]

Many Californians felt that the sometimes uncoordinated and often scattered work on air pollution had proven to be inadequate to cope with the smog problem; a new approach was needed.

BIRTH OF PROJECT CLEAN AIR

At the dawn of the 1970s, a new surge in public pressure for accelerated smog control programs came from dissatisfaction with policies which had failed to relieve the urban areas of California—and, increasingly, nearby

rural areas which began to become polluted—of smog. Political activity on
environmental issues was at a high level as legislators were enjoying rides
on the ecology bandwagon. To the experienced observer of California poli-
tics it became apparent that the time was ripe to pluck more money from
the state in the search for quicker solutions to air pollution problems. It
was in this atmosphere of demands for action on the one hand, and feelings
of anticipation that the state would move more strongly to support needed
research in air pollution on the other, that Project Clean Air was born.

In November 1969, UC President Charles Hitch wrote to the Chancellors
of the nine campuses advising them that the University was launching "a
new and aggressive approach to mission-oriented problem-solving effort on
a large scale [directed at the] elimination of smog and other air pollution. . . ." Project Clean Air was thus conceived.

Plan of Action

In the single month that followed, a plan of action was written by Riverside campus personnel working with a special nine-campus committee appointed by President Hitch. Stahrl Edmunds, Dean of the UC Riverside
School of Administration, was selected as chairman of the committee by
Chancellor Ivan Hinderaker. Faculty members were invited to submit their
suggestions for mission-oriented research; 120 proposals were received. The
plan was developed from these proposals and the committee deliberations.
It was formally presented to key California legislators on the opening day
of the new Legislature, 2 January 1970, and the Board of Regents reviewed
the same document on 22 January. Reaction from both bodies was mixed:
The basic theme of a new mission-oriented air pollution research program
was judged relevant, but the details of the plan were considered inadequate.
The Regents were gratified with the effort to date and loaned funds ($150,-
000) to support further planning efforts.

University-wide Planning Conference

Project Clean Air was only partially formed at this point. The charge
from the Regents was to produce a more detailed proposal with specific
research goals, time schedules, and budgets. An unprecedented, multidisciplinary University-wide faculty conference was held in February 1970 to
take apart the original proposal, reshape it, and put it back together again.
This was successfully done through the use of study groups. Each of the
fields of air pollution research was analyzed by a faculty group during the
three-day conference. The reports written by these study groups provided
a composite picture of what Project Clean Air should be. There was no
precedent to refer to since there was no other all-campus, multidisciplinary
program in the University system. The forty conferees further agreed on

the basic organization and management structure of the Project. It was emphatically urged by the conference that the top man in Project Clean Air be a top man—administratively speaking—in the University, reporting directly to the President. The faculty felt that only with a high-level commitment on the part of the University-wide administration could this unique program be successful.

At about this same time the All-University Faculty Conference was considering a report by a study committee on the "Role of the University in Problem Solving for the Larger Community." The study committee had wrestled with the general problem of shaping policies for managing the research capabilities of the faculty around urban and environmental problems. It drew heavily from the concurrent work of structuring Project Clean Air. In support of the manner in which Project Clean Air was organized, the report says: "[It was] suggested that scholars from different schools or departments regroup themselves around policy issues in temporary research units. Certain elements of this idea may be found incorporated in the recently proposed [Project Clean Air] program for assisting California in control of its air pollution problem." [3]

Revised Proposal

The ideas and study group reports accumulated at this conference were the basis of the revised proposal which was submitted to the Regents in March 1970. This new document spelled out short-term (sixteen–eighteen month) and medium-term (eighteen–thirty-six month) research as recommended in the study group reports. Research projects were divided into seven categories: (1) automobile engine development; (2) human health effect; (3) social sciences; (4) meteorology and simulation modeling; (5) power, industry, agriculture and ecology; (6) instrumentation development; and (7) atmospheric chemistry and physics.

The selection of these categories relied on the expressed research interests of faculty members and the recommendations of the study committees. The revised proposal was acknowledged to be only a skeleton which needed further definition by the faculty. Seven Task Forces were proposed, therefore, to hammer out the missing parts of the proposal. The Task Forces would be extensions of the study groups; their task was to assess the state-of-the-art of air pollution control in California and recommend specific applied research topics to help solve the most pressing problems in each category.

The proposal reflected the thinking of the study group report:

Clearly, the University is not in a position to actually *solve* any of the critical problems facing our society. Its role must be to inform decision makers and the general public about the existence of problems which need solutions and to recommend alternative ways of dealing with them. Improvements can be brought about only through the action of those public and private decision makers who are vested with the authority and the responsibility to act.

The report also cautioned:

Any effort to expand the University's role in problem solving for the larger community must be cognizant of the risks involved.[3]

The Project Clean Air proposal, carrying a total request for 16,470,000 dollars in a three-year mission-oriented research program, was formally approved by the Regents in March 1970. Additional funds of 50,000 dollars were advanced to commence the Task Force Assessments, and an Executive Director was appointed. This completed the birth of Project Clean Air.

The University's three original major goals or objectives for Project Clean Air remain essentially unchanged:

1. Act as an evaluator of technical claims and serve the state government, particularly the Air Resources Board, the State Department of Public Health, and the State Legislature, by providing knowledge and recommendations for legislative decisions on air quality control
2. Organize the University's interdisciplinary capabilities in pertinent scientific, technological, economic, sociological, and medical areas and direct them in an integrated effort to reduce air pollution
3. Provide technical assistance to keep state and local agencies informed and aid them with problems of training, implementation, and enforcement

Six-month Experiment

With some money and a great deal of enthusiasm Project Clean Air began a probationary existence in April 1970. It was probationary because the 16 million dollars needed to support the program would have to be generated by new state legislation; it was important that the University perform well with the limited amount of funds loaned by the Regents if it hoped to be made the custodian of a large and unprecedented state-authorized research fund.

Task Force Assessments

The assessment process was started quickly. The February study groups, with some modifications, became the seven Task Forces. Typically, a small group of professors (from two to twelve) was constituted as a Task Force; this was augmented by graduate and undergraduate students in some cases. Meetings were held to plan how to attack the problem of making a comprehensive technical assessment of the assigned air pollution research field. Some groups held workshops inviting experts in the field from inside and outside the University to give brief papers on their research activities. As the state-of-the-art became defined and the assessment document became structured, elements of the assessment document were assigned to individual task force members. These various elements were drafted, again with the assistance of outside experts, and then were edited by the Task Force Chairmen. All this discussion, drafting, and editing took three months to accomplish. The final draft assessment documents were presented for review to a University-wide conference of 150 faculty members, which was held in late July 1970. Comments and criticisms from faculty members were incorporated into the reports, and formal publications were released in September.

On balance, the Task Force Assessments were—and remain—the most comprehensive assessment of the state of air pollution control in California.[3] The faculty and staff—over 100 persons were involved in preparing the assessment documents—delivered on the promise to thoroughly study the air pollution problems of the state and make recommendations for needed applied research. These reports formed the basis of the University's planned air pollution research program. But it was destined to be modified by political actions that had been operating concurrently with the assessment process during the spring of 1970.

Pushing Legislation

The political events pertinent to Project Clean Air formally started in April 1970 with the introduction of a bill calling for a 16 million dollar research program. The roots of this action trace back to January of that year when early informal discussions about Project Clean Air had resulted in Senate-sponsored legislation which was written around the University's proposed program. Other bills were also introduced to accomplish similar things; but only the Senate Bill 848 was successful over all the legislative hurdles in Sacramento.

During the April–September period when PCA Task Forces were hard at work planning the last details of the elaborate research program, the Executive Director was laboring to keep the program alive by working to insure passage of SB 848. The bill was actually defeated in committee on more than one occasion; it was revived by a combination of legislative maneuvering by the sponsor and technical briefings by University representatives. This work was not done without some hesitation. It was becoming increasingly obvious that the University was *not* going to become the recipient of a generous air pollution fund; it was also clear that the passage of the bill at all was in doubt. But the persistent effort was rewarded on 20 September 1970, when Governor Ronald Reagan signed SB 848 into law.

The fault in the law, as far as the University was concerned, was that only 9 million dollars was allocated for applied research—not 16 million dollars. But more seriously, the money was placed under the control of the Air Resources Board—not the University of California. There was concern on the part of University officials whether Project Clean Air could function as originally planned under these state-imposed restrictions. The law did say that 750,000 dollars should be allocated *immediately* to the University for the development of its research program. There was some hope for the future of Project Clean Air.

Short-term Research

A third phase of Project Clean Air, during the six-month probationary period, was the performance of some actual research. Back in April some unexpended state ("urban crisis") funds in the amount of 187,000 dollars were diverted to Project Clean Air; with this money twenty-three short-term research projects were conducted from 1 April through 30 June 1970. This unexpected windfall of state money gave the University a chance to show a capability to perform truly short-term applied research on air pollution problems. Five key accomplishments of this three-month program were:

1. Seawater was successfully used to scrub sulfur dioxide from industrial stack gases with 99 percent efficiency—a process with a potential national savings of 200 million dollars annually.
2. A benefit-cost analysis of control devices for used cars pointed out the need for different devices for different ages of cars and the importance of "clean" fuels in reaching "acceptable" emission levels.
3. Oxidant levels in Los Angeles were correlated with motor vehicle accident rates which pointed to the necessity for further study of the "safety hazards" of photochemical smog.
4. Aerial studies of the inversion layer between Santa Monica and Riverside invalidated previous over-simplified concepts of how pollutants are trapped in the South Coast Air Basin; unexpectedly high concentrations of smog were found in layers above and below the inversion boundary, with alternating layers of clean air.
5. Exposure of animals to ozone caused an interference with the normal mechanisms used to defend against inhaled bacteria, suggesting that urban human population may be subject to an increased risk of respiratory infections as a consequence of photochemical smog.

Project Clean Air Research Reports on the results of the twenty-three projects were compiled and published in September 1970 as a companion set of four volumes to the Task Force Assessments.[4]

CURRENT OPERATIONS

Springing the Money

It was October 1970. The Governor had signed SB 848. The Project Clean Air Assessments and Research Reports were off the press. The six-month experiment had ended. The Regents had heard a progress report and were sufficiently impressed with the accomplishments of Project Clean Air to advance more money, another 100,000 dollars, to put the program into full gear without waiting for the state appropriations to materialize. Project Clean Air was, or so it seemed, a full fledged University-wide program.

The first task of the newly assembled operation in the President's office in Berkeley was to obtain release of the money—the 750,000 dollars that was to come "immediately" to the University. Informal contacts gave the first clues to the manner of interpreting "immediately": The University would first have to submit a detailed proposal of how it intended to spend the 750,000 dollars. This was done by drastically revising the past proposal documents. Revision was necessary because legislators' desires about the nature of appropriate air pollution research had to be taken into account. During the summer battle to pass the bill, impressions were accumulated—some were written into the legislation and some were expressed only verbally—relative to what research the Legislature would and would not approve. One result of this was that social science research was specifically deleted from any listing of applied research. The carefully laid plans of the University economists, political scientists, sociologists, and others had to be largely set aside because of the political reshaping of the research program.

A Research Plan, carefully modified to reflect the "new" Project Clean Air, was submitted to the ARB in late October. In mid-November word was received by the President's office that the 750,000 dollar allocation had been released to the University. It looked as though "immediately" would be two months or so. But while verbal approval of the research plan had been received, the first delaying tactic took place: The ARB requested that the University submit a detailed list of research projects before approval to spend funds could be given. "Immediately" suddenly became undefinable.

Proposal Mill

Any research scientist dependent on funds from an outside agency knows what came next. November and December were the months of the proposal mill. In order to satisfy the ARB's request, it became necessary to solicit formal proposals from the faculty. A year's accumulation of proposals was on hand in the Project Clean Air files, but these were out-of-date and not in a form ready to be presented to the state agency. A call went to interested faculty members to bring their proposals up-to-date and submit them for faculty review. The need for faculty review was considered to be a vital part of Project Clean Air's operation; proposals to be sent to the state agency should be the best and most pertinent work, and this could be insured only with a thorough evaluation by reviewing committees. The administrative procedures were established in the University-wide office, and the review and selection of the top thirty-eight proposals were selected for transmission to the ARB. This package went forward on 8 January 1971. It was anticipated that approval to spend the 750,000 dollars would be given soon. "Immediately" looked like four months at the most. But again came the frustration of further delay. The word came: The ARB would require two months to review the proposals.

How to Live With a Bureaucracy and Learn to Like It

Four months had passed and none of the state money had been spent on air pollution research as directed by SB 848. Another two months would be used by the ARB to resift the already sifted proposals, and thus duplicate the administrative effort undertaken by Project Clean Air. In effect the state had stripped away most of the duties assumed by the PCA office. This action signalled the end of the originally intended involvement for the University in the state-sponsored program.

In retrospect the repeated delays and the gradual assumption of control by the ARB occurred in the normal process of establishing a new state program. Over a period of three months, the ARB hired the staff to administer the new research fund; it was in first gear, administratively speaking.

The state procedures finally ran their course. On 18 March 1970, six months after the enabling legislation was signed into law, the Air Resources Board announced the allocation of 1.1 million dollars of the 9.25 million dollar research funds. The initial list of nineteen approved research projects included fifteen University proposals. The 925,000 dollars awarded to UC scientists was 85 percent of the funds released.

The state had received and reviewed 135 proposals; one-half of these were from the University (including those rejected by the PCA reviewing committees) and half were from private corporations, research organizations, and other institutions. Sixty-eight proposals were rejected, and forty-eight received some preliminary approval. Most of this latter group were subdivided into research categories needing further definition. Authors would be invited to a series of topical conferences to discuss research needs; following this, requests for proposals would be prepared by the ARB staff and new proposals received and reviewed—all in an effort to sharpen the proposals to meet present state needs. It appears certain that the ARB will refine this conference request for proposal technique, and continue to use it in the future development of its research program.

Concurrent with the ARB's March announcement of awards, was the University's announcement from the office of the President that it was phasing out Project Clean Air. The reasons given for the change centered upon the difficulties in bringing to reality the University-wide coordinated research program originally proposed as Project Clean Air. The delays in the release of funds and the duplication of administrative procedures by the ARB were cited as specific causes for de-emphasizing the program.

At this writing the final impact of these recent events cannot be measured. Clearly the University and the state have acted independently; they thus far are short of a common ground of understanding and development of complimentary methods of operation. Under these circumstances, the University has elected to go into low profile and allow the ARB to continue the management of the program over which it has ultimate financial control.

THE FUTURE

Project Clean Air has had an exciting history. Originating a University-wide, multi-disciplinary research program around a politically and socially important problem of contemporary society proved to be a great challenge. Unfortunately, the problems of carrying out the administrative actions became so large they almost engulfed and destroyed the original concept. But the entire process was not without its successes and deserves much better than to be labeled a failure. This is true in part because the story of Project Clean Air is still being written.

In California the priority of the air pollution issue is slowly rising, but no drastic change of its position on the priority list can be expected. Surely, it will continue to be a major issue in the state. It would be unrealistic to expect air pollution to suddenly change its position of priority among social ills. Unforeseen political pressures could cause this to happen, however. The present prominence of the problem should be appreciated and full advantage should be taken of the political climate that now favors strong action.

The amount of money being directed to air pollution by the state is rising, with some possibility of even more funds becoming available from the diversion of present fuel taxes or creation of new taxes. So while the sense of urgency may not change much in the next few years, the amount of money may, making possible more of the research needed to help solve the problem. If bills to add funds to air pollution programs now before the Legislature fail this year, perhaps new sources of support can be created later, in time to maintain at least a constant level of funding. It appears likely, therefore, that California will continue to support some significant level of air pollution research for the first half of this decade, at least.

It is impossible to predict the details of the role Project Clean Air will take in the future, as these are now being negotiated. However, some wide limits within which the Project will probably operate can be suggested. Given a relatively high priority in California for air pollution problems and some continuing state funds for research that is needed to help solve the problems, it is certain that the University of California will continue to play a major role in air pollution research. The University's past performance should demonstrate to state government officials that it possesses real capabilities that should be utilized to their fullest extent to enhance and support California's attack on air pollution.

The University of California has had a major role in California's air pollution research and control activities before and since Project Clean Air; it will continue in this role with or without Project Clean Air. If nothing else, Project Clean Air has served the purpose of focusing attention on the continuing need for scientific research to support legislation in the technically complex field of air pollution. An unprecedented allocation of state funds was the result. Project Clean Air has been a totally new and unique experience.

All the rules and conventions of "grantsmanship" were set aside as a group of skilled and dedicated faculty and staff combined their talents and energies to develop an unprecedented state-sponsored research program. It did not fail; it just did not achieve 100 percent success. On balance the successes outweigh the disappointments. The University is proud of Project Clean Air, and will continue to seek ways to put its research and administrative talents to work on relevant problems of contemporary society.

Reference notes

1. A. J. Haagen-Smit et al., "Investigation on Injury to Plants from Air Pollution in the Los Angeles Area," *Plant Physiology,* January 1952.
2. John T. Middleton, "Air Pollution: Where Are We Going?" (Presented to a conference on environmental pollution sponsored by the Junior League of Los Angeles and the Rand Corporation, Los Angeles, Calif., 6 December 1969).
3. University of California, *Project Clean Air Task Force Assessment,* 4 vols., (1 September 1970).
4. University of California, *Project Clean Air Research Reports,* 4 vols., (1 September 1970).

19. National Goals

J. Frederick Weinhold
Office of Science and Technology

In the light of energy and environmental needs, let us examine the extensive work on electric power plant siting, performed by the Office of Science and Technology, which can provide a basis for suggesting some national goals in this area.

The Office of Science and Technology is a small group of senior scientists, engineers, and others within the Executive Office of the President under the direction of President Nixon's Science Advisor, Dr. Edward E. David, Jr. The Office, established almost ten years ago, is primarily an advisory and coordinating organization involved in research and development policy and other issues involving technology. About three and one-half years ago a small group specifically concerned with energy policy was established within the Office to deal not only with research and development, but also the broad range of economic and institutional problems involving energy.

ELECTRIC POWER PLANT SITING

One of the first major projects of the energy policy group involved electric power plant siting. In late 1967 and 1968, a few individuals began to recognize the objections being raised to nuclear power plants, in California and along the east coast as well, and to a few other electric power facilities, as symptoms of a much larger impending problem. The study involved the work of several federal agencies; therefore, an interagency study group was established to look first at the factual basis of the problem and later to recommend policy changes to help solve the problem. This group included the Atomic Energy Commission, the Department of the

341

Interior, the Federal Power Commission, the Air Pollution Control Office now in the Environmental Protection Agency, the Rural Electrification Agency, the Tennessee Valley Authority, and the Council on Environmental Quality after it was formed. The first report, issued in January 1969, entitled "Considerations Affecting Steam Power Plant Site Selections," attempted to put into perspective environmental problems associated with nuclear plants, fossil-fuel plants, and transmission lines. In the course of this effort, the similarities between nuclear and fossil plants became apparent, as did the need for improved institutional arrangements and technology to deal with the problems.

Early in the Nixon Administration, the study group was reconvened to deal with the policy questions and met with representatives of conservation groups, state and local government, and all segments of the electric utility industry. Although the specifics differed, it was found that power plant siting and transmission line routing problems existed all over the country, not just in the populous coastal areas where they had been highly publicized. During the study, degradation of the environment became a popular national concern. Numerous bills were introduced in Congress and state legislatures, while court suits, seeking to halt construction or operation of new plants, became commonplace.

In this context the study group developed its basic ideas and recommendations, which were issued last fall in a report entitled "Electric Power and the Environment." The fundamental concept of the report was that industry, in this case the utilities, must retain the primary responsibility for planning and building needed new facilities. However, the public, primarily through cognizant government agencies, must play an expanded, meaningful role in the decision-making process. In essence, a decision-making body must exist which can decide from an overall public interest point of view: whether or not the new facility is needed, and, if so, is proper regard being given to environmental concerns?

The key to this process is the need for balanced decision making, which should also be adopted as a national goal. Environmental quality and energy are both important national concerns in their own right. Often these concerns become interrelated, as in the specific case where the construction of a new power facility required for system reliability will impair environmental quality or vice versa. At the present time, there are generally no institutional arrangements which can decide this broad question. Nearly all of the current hearings and procedures, and there are a great many of them, focus on a single aspect of the problem, such as nuclear health and safety, air pollutants, thermal discharges, or financial questions. Environmentalists raising the fundamental issue usually have no forum other than the press in which to voice their concerns.

This question of balance is a difficult concept, but I think David Sive, a noted environmental lawyer, aptly described it in his 1971 address to the National Academy of Engineering Forum on Power Plant Siting. He suggested that a regulatory agency reviewing a proposal for a new power plant can begin with one of three presumptions: (1) We need the additional power—now how can we best fit the plant into the environment? (2) We need to protect the environment—now can it tolerate a new power plant? or (3) Recognizing environmental protection and more power as both important goals—what is the best combination in this particular situation? A truly balanced approach requires the latter presumption and must involve the subjects of overall power growth rates as well as land use, air, water, and solids pollution.

In February 1971, President Nixon proposed in his Environmental Message legislation to implement the new institutional arrangements recommended in the report, "Considerations Affecting Steam Power Plant Site Selections," and, on 10 February 1971, Dr. David transmitted the specific legislative proposal to Congress. This proposal was the subject of hearings by the House Commerce Committee beginning 4 May. The bill requires open long-range planning for all bulk power facilities at least ten years in advance of their construction, designation and approval of alternative power plant sites five years in advance of construction, and specific licensing of new power plants and transmission lines two years prior to construction.

The study group concluded that this action was best carried out at the state level or, if the states chose, at the regional level. It was felt that the problems were too large to be resolved at the local level, but diverse enough to require a nonfederal approach. The proposed legislation, therefore, encourages each state either to establish its own power plant certifying agency or to join a regional agency. As in the case of air and water pollution control, the federal government stands ready to do the job if the states choose not to.

A certifying agency is really the key to balancing energy needs and environmental protection since its members will be required to review the long-range plans, to approve the alternative sites, and, finally, to certify each large power plant and transmission line before construction can actually begin. The draft legislation permits the states a good deal of flexibility in the establishment of these agencies, but the law itself and subsequent detailed guidelines to be promulgated by the president spell out who must participate in the decision-making process and the relationship of the certifying agency's decision-making body to other state and local agency requirements. To facilitate the construction of new facilities, once it is decided that they should be built, an attempt is being made to make the state certification procedure as nearly "one stop" as possible.

All bulk power facilities will be included under the proposed legislation regardless of whether they are publicly or privately owned. A federal certifying agency will therefore be designated by the president to certify federally owned plants and those in states which do not have an approved certifying body. Again, we are striving to make the federal procedures as nearly "one stop" as possible to facilitate construction once the decision is made. The draft even provides for the use of federal eminent domain when needed if a state, regional, or federal certificate of site and facility has been obtained.

A key ingredient to balanced decision making is participation by all the interested parties in the process, but another is consideration of all the questions which involve the public interest. The most obvious ones are air and water pollution controls. Is the plant designed to meet applicable standards and minimize any undesirable effluents? Once standards have been adopted, this process is primarily one of engineering and economics. Policy judgments will be needed to decide how much should be spent, if any, to go beyond present standards. It is envisioned that most of these questions will be resolved at the preconstruction certification proceedings.

Land use is another crucial question which is not yet as well recognized or formulated as control of air and water pollution, however it is at the base of most controversies today. Support is readily obtained for more electric power produced with minimal pollution, but when it comes to siting the plant next to residential, recreational, or natural areas strong opposition is found. Regardless of technology—even if, for example, it were feasible to bury the power plants completely underground—there would still be some objection because construction and operational access ways would undoubtedly affect the environment. Making decisions on land use is therefore more a political problem than an engineering one. It is necessary to balance the harm to some individuals who do not want a power facility in a particular location with the benefits to those who do and those who want the power no matter where it is produced. Land use questions will be the predominant issues at the five-year decision point when alternative sites are selected.

Finally, the power needs terms in the balance equation must be evaluated. The rate of electric power growth is an issue which most traditional utility executives attempt to avoid, but it is the most important question for some environmentalists and a question increasingly being raised by public officials. Many of the laws passed a generation or so ago actively promoted the growth of low cost, abundant supplies of power. It was almost an article of faith. As the traditional 7 percent growth in electric power consumption is projected to the end of the century, however, serious environmental problems would result, such as, large amounts of land being used for transmission line routes, great quantities of water being evaporated in the process of cooling steam plant condensers, and huge amounts of air pollutants and radioactive wastes being produced. The rapid use of our non-renewable fuel resources is also a point of concern. As will be discussed later, new technology offers the best long-range hope for resolving this dilemma, but the long lead time, uncertainty of the results, and the serious consequences of failure require that we consider measures to control growth as well.

REDUCING POWER GROWTH RATES

There are two different approaches to reducing power growth rates: (1) increasing the prices to reflect true costs so that consumers will not purchase as much electricity or (2) improving the efficiency of consumption. The first is the economist's or rate maker's approach, while the second is the engineer's; however, in reality, the best solution involves both. If rates are increased sufficiently to cover all of the environmental costs of power production and if rates are redesigned to eliminate promotional practices, there will be real economic incentives for engineers to build more efficient appliances, industrial processes, and heating and cooling systems. There is a need for consumer education to ensure that the benefits are understood and more efficient energy systems are demanded; in addition, there exists a need for similar policies across the energy spectrum to prevent undesirable switches to other energy forms. Today only the most sophisticated commercial and industrial consumers make decisions based on minimum life cycle costs rather than lowest initial cost. Recognizing this or perhaps ensuring it, manufacturers and builders design for minimum initial cost and seldom give ordinary consumers a real choice.

Promotional rate structures—special low rates per kilowatt-hour for certain large industrial and other consumers—undoubtedly made sense when we were trying to expand the economy in a particular region and reduce overall rates through economies of scale. Today these policies must be reviewed when the issue is now one of using up our natural resources too rapidly or building new high cost capacity to meet new loads which will raise rather than lower average rates. In the long run, reducing the growth rate of power consumption will not solve the conflict between energy and environmental needs, but it will give us one very valuable commodity: time. Wise use of this time may produce new technologies through research and development which would result in having both the energy and the environment we desire. Recognizing this problem, the longest chapter in the report "Electric Power and the Environment" was devoted to the new technologies needed to resolve electric power problems.

AIR POLLUTION CONTROL

In the estimation of many, the most urgent needs for new technology involve the controlling of pollution from power plants, particularly the air pollution from fossil-fuel plants. No processes have yet been adequately demonstrated on a commercial scale for controlling sulfur oxides from the burning coal or oil. A number of systems are in the development stage, and some are currently being demonstrated on a scale large enough to demonstrate their commercial applicability. The utilities and government want to ensure that if large amounts of capital are spent to install this equipment, it will work reliably, reducing the SO_2 concentration in stack effluents to prescribed levels, and not jeopardize the operation of the unit.

Since power plants are large, the pollution control equipment is large and expensive. Demonstrations, therefore, cost on the order of 10 to 15 million dollars or more for each process. Both government and industry are involved in this effort, jointly funding some of the more promising approaches. Stack gas cleaning is perhaps the approach drawing the most attention. Demonstrations are underway on wet and dry limestone processes, catalytic oxidation, and magnesium oxide scrubbing processes. A recent NAE report details many of the other processes currently in the research and development stage.

In addition to these stack gas processes, coal cleaning and new combustion techniques also offer some promise. Work on the other major air pollutants, nitrogen oxides and very fine particulates, however, is just beginning.

Controlling and utilizing waste heat is yet another major power plant pollution problem requiring research and development studies. The generation of electric power by either nuclear or fossil fuels produces large quantities of waste heat which is now frequently dumped into adjacent natural waterways with potentially serious physical, chemical, and biological effects. Wet cooling towers eliminate some of these environmental problems but precipitate new ones, such as area fogging and synergistic health effects from combustion air pollutants. Heat rejection is a problem where large supplies of freshwater are not available. Dry cooling is generally an untried approach because present design concepts are costly and reduce the efficiency of power generation. Perhaps the most intriguing prospects for waste heat disposal involve the development of beneficial uses for the heat, in essence, converting a pollutant into an asset.

There is one other environmental problem associated with electric power that must be solved in the next few years: the development of low cost, underground, high voltage transmission technology. Recent work on electric power plant siting problems showed that delivering electricity to the customer frequently raises as many environmental objections as generating it. Underground transmission offers a means of reducing the land required and eliminating much of the opposition, if technical and economic problems can be overcome. A number of different proposals are in the research phase, ranging from cost-cutting improvements of present technology to supercon-ducting lines capable of handling thousands of megawatts of power.

NEW POWER SOURCES

In addition to these near-term pollution control technologies, several new power generation systems offer the promise of conserving valuable re-sources and reducing pollution in future decades. The development of new high efficiency fossil-fuel power conversion systems, such as magnetohy-drodynamics (MHD) and combined gas turbine–steam turbine cycles, offer the possibility of reducing both the environmental problems and the fuel required to produce electric power. Both systems require development, however, and it will be a number of years before they provide significant amounts of power. Fossil fuels will be important in electric power genera-tion for years to come, primarily because of their lower capital costs and flexibility. Cycles to use them more efficiently are needed.

Advanced nuclear power systems are also prime candidates for use in the decades ahead. It has long been recognized that for nuclear fission to serve the long-term energy needs of the nation and world, a breeder reactor utilizing a high proportion of the energy available in natural uranium was needed. Currently, large programs here and abroad are concentrating on the liquid metal fast breeder (LMFBR), but additional work remains before the system becomes commercial. The molten salt and fast gas concepts offer alternative approaches, should it become desirable or necessary to expand the program.

Although there are many scientific problems still to overcome and a long period of engineering development ahead, nuclear fusion seems particularly attractive because of fuel availability and environmental compatibility. The idea of using the virtually unlimited supplies of heavy water in the sea for energy has challenged scientists for two decades.

Perhaps even farther in the future is solar power. The sun provides a great renewable source of energy, but the technology is not available to use it in significant quantities. A variety of ideas utilizing satellites to supply electricity have recently been postulated. Since little or no waste products are produced, it is a particularly acceptable method from an environmental point of view.

On earth, the fuel cell, organic Rankine cycle, and other systems have been suggested as sources of low pollution energy for small scale on-site power generation. The potential efficiency of such systems offers real advantages in certain locations, particularly where the waste heat can be used for space heating, cooling, or other requirements.

Several other nonconventional energy sources, such as tidal and geothermal power, have been advanced by scientists, and it is possible that developments in these fields may lead to new sources of clean energy.

The cost of the research and development programs to develop complete new power generation technologies is high, frequently reaching to the hundreds of millions or occasionally billions of dollars. Even the largest private firms cannot afford the financial risks involved, but since the technology will be used by private industry they must be financially involved. New methods of funding and government-industry management will be required.

OTHER ENVIRONMENT VS. ENERGY CONFLICTS

It is always somewhat presumptuous to generalize from the specific area in which one has worked most closely; however, I believe the elements of the study group's proposed solution to the conflict between electric power and environmental needs are applicable to other energy forms.

New institutional mechanisms are needed to balance national interests involving other energy related problems, such as oil and gas leases, strip mining, pipelines, oil refineries, and terminal location. President Nixon's proposal for a Department of Natural Resources is a vital first step toward putting the federal energy establishment in order to deal with these questions in a broad-based manner.

The need to control the growth in energy demand is a general problem. Natural gas consumption, for example, has been growing nearly as fast as electric demand, while its resource base seems to be quite limited. Many expect that domestic natural gas production will peak within the next two decades. Coal resources are quite large in comparison, but oil resources are of the same order as natural gas resources. The key is to learn how to use each of these fuels more efficiently and in conformity with environmental standards.

Technology provides a key to both of the aforementioned, through increased consumption efficiency and new approaches which will permit shale and coal reserves to be converted into clean, convenient fuels. Coal gasification, oil shale recovery, and nuclear stimulation of natural gas formations are examples of methodologies which will be utilized in the years ahead.

In summary, I suggest that we need to set three separate but related goals in order to meet environmental as well as energy needs in the decades to come. These include: (1) Establishing new institutional mechanisms which will facilitate the balancing of environmental and energy needs on specific projects in specific locations; (2) Learning how to control energy growth through increasing the efficiency of energy utilization; and (3) Gaining the necessary time to develop new technologies which will solve the problems of energy and environmental needs, through adoption of these measures.